Lecture Notes in Computer Science   11044

Commenced Publication in 1973
Founding and Former Series Editors:
Gerhard Goos, Juris Hartmanis, and Jan van Leeuwen

More information about this series at http://www.springer.com/series/7412

Danail Stoyanov · Zeike Taylor
Enzo Ferrante · Adrian V. Dalca et al. (Eds.)

Graphs in Biomedical Image Analysis *and* Integrating Medical Imaging and Non-Imaging Modalities

Second International Workshop, GRAIL 2018
and First International Workshop, Beyond MIC 2018
Held in Conjunction with MICCAI 2018
Granada, Spain, September 20, 2018
Proceedings

 Springer

Editors
Danail Stoyanov
University College London
London
UK

Enzo Ferrante ⓘ
CONICET/Universidad Nacional del Litoral
Santa Fe
Argentina

Zeike Taylor
University of Leeds
Leeds
UK

Adrian V. Dalca ⓘ
Harvard Medical School
Cambridge, MA
USA

Additional Workshop Editors *see next page*

ISSN 0302-9743   ISSN 1611-3349   (electronic)
Lecture Notes in Computer Science
ISBN 978-3-030-00688-4   ISBN 978-3-030-00689-1   (eBook)
https://doi.org/10.1007/978-3-030-00689-1

Library of Congress Control Number: 2018954664

LNCS Sublibrary: SL6 – Image Processing, Computer Vision, Pattern Recognition, and Graphics

This Springer imprint is published by the registered company Springer Nature Switzerland AG
The registered company address is: Gewerbestrasse 11, 6330 Cham, Switzerland

Additional Workshop Editors

Tutorial and Educational Chair

Anne Martel
University of Toronto
Toronto, ON
Canada

Workshop and Challenge Co-chair

Lena Maier-Hein
German Cancer Research Center
Heidelberg
Germany

Graphs in Biomedical Image Analysis, GRAIL 2018

Sarah Parisot
AimBrain
London,
UK

Bartlomiej Papiez
University of Oxford
Oxford,
UK

Aristeidis Sotiras ⓘ
University of Pennsylvania
Philadelphia, PA
USA

Integrating Medical Imaging and Non-imaging Modalities, Beyond MIC 2018

Mert R. Sabuncu ⓘ
Cornell University
Ithaca, NY
USA

Li Shen ⓘ
University of Pennsylvania
Philadelphia, PA
USA

GRAIL 2018 Preface

GRAIL 2018 was the 2nd International Workshop on Graphs in Biomedical Image Analysis, organized as a satellite event of the 21st International Conference on Medical Image Computing and Computer Assisted Intervention (MICCAI 2018) in Granada, Spain. After the success and positive feedback obtained last year, this was the second time we brought GRAIL to MICCAI, in the spirit of strengthening the links between graphs and biomedical imaging.

The workshop provides a unique opportunity to meet and discuss both theoretical advances in graphical methods, as well as the practicality of such methods when applied to complex biomedical imaging problems. Simultaneously, the workshop seeks to be an interface to foster future interdisciplinary research including signal processing and machine learning on graphs.

Graphs and related graph-based modelling have attracted significant research focus as they enable us to represent complex data and their interactions in a perceptually meaningful way. With the advent of Big Data in the medical imaging community, the relevance of graphs as a means to represent data sampled from irregular and non-Euclidean domains is increasing, together with the development of new inference and learning methods that operate on such structures. There is a wide range of well-established and emerging biomedical imaging problems that can benefit from these advances; we believe that the research presented in this volume constitutes a clear example of that.

The GRAIL 2018 proceedings contain 5 high-quality papers of 8 to 11 pages that were pre-selected through a rigorous peer review process. All submissions were peer reviewed through a double-blind process by at least 2 members of the Program Committee, comprising 18 experts in the field of graphs in biomedical image analysis. The accepted manuscripts cover a wide set of graph based medical image analysis methods and applications, including neuroimaging and brain connectivity, graph matching algorithms, graphical models for image segmentation, brain modeling through neuronal networks and deep learning models based on graph convolutions. In addition to the papers presented in this LNCS volume, the workshop comprised short abstracts and two keynote presentations from world-renowned experts: Prof. Michael Bronstein and Prof. Dimitri Van De Ville. We hope this event will foster the development of more powerful graph-based models for the analysis of biomedical images.

We wish to thank all the GRAIL 2018 authors for their participation and the members of the Program Committee for their feedback and commitment to the workshop. We are very grateful to our sponsors Entelai (https://entelai.com/) and the UK EPSRC-funded Medical Image Analysis Network (MedIAN - https://www.median.ac.uk/) for their valuable support.

The proceedings of the workshop are published as a joint LNCS volume alongside other satellite events organized in conjunction with MICCAI. In addition to the LNCS volume, to promote transparency, the papers' reviews and preprints are publicly available on the workshop website (http://grail-miccai.github.io/) together with their corresponding optional response to reviewers. In addition to the papers, abstracts, slides, and posters presented during the workshop will be made publicly available on the GRAIL website.

August 2018

Enzo Ferrante
Sarah Parisot
Aristeidis Sotiras
Bartlomiej Papiez

Organization

Organizing Committee

Enzo Ferrante   CONICET and Universidad Nacional del Litoral, Argentina
Sarah Parisot   AimBrain, UK
Aristeidis Sotiras   University of Pennsylvania, USA
Bartłomiej Papież   University of Oxford, UK

Scientific Committee

Kayhan Batmanghelich   University of Pittsburgh and Carnegie Mellon University, USA
Eugene Belilovsky   Inria, France and KU Leuven, Belgium
Siddhartha Chandra   Inria, CentraleSupélec, France
Xin Chen   The University of Nottingham, UK
Emilie Chouzenoux   Inria, CentraleSupélec, France
Puneet K. Dokania   Oxford University, UK
Ben Glocker   Imperial College London, UK
Ali Gooya   University of Sheffield, UK
Mattias Heinrich   University of Luebeck, Germany
Lisa Koch   ETH Zurich, Switzerland
Evgenios Kornaropoulos   University of Cambridge, UK
Sofia Ira Ktena   Imperial College London, UK
Georg Langs   University of Vienna, Austria and MIT, USA
Jose Ignacio Orlando   Medical University of Vienna, Austria
Yusuf Osmanlioglu   University of Pennsylvania, USA
Yangming Ou   Harvard University, USA
Nikos Paragios   Inria, CentraleSupélec, France
Sotirios Tsaftaris   University of Edinburgh, UK
Maria Vakalopoulou   Inria, CentraleSupélec, France
William Wells III   Harvard Medical School, USA

Sponsors

Beyond MIC 2018 Preface

Beyond MIC (http://beyondmic.mit.edu) is a full-day workshop on integrating medical imaging and non-imaging modalities to answer novel clinical and healthcare challenges. Recent large-scale, multi-site data collection efforts - such as the ADNI, TCIA, and UK Biobank - and the hospital open-data initiatives are resulting in growing datasets of medical images. Increasingly, these studies often also include non-imaging modalities such as electronic health records, insurance data, pathology reports, laboratory tests, and genomic data. Alongside medical images, these rich external sources of information present an opportunity for improving traditional medical image computing tasks like diagnosis, prediction, and segmentation, as well as facilitate newer tasks like automatic image annotation. However, these heterogeneous data also poses unprecedented technical obstacles, including pre-processing different or inconsistent formats, modeling the complex noise and heterogeneity, jointly handling high dimensionality and multi-modality, and optimizing computational resources to handle significantly larger amounts of data.

Machine learning methods tackling data-driven health care problems have been gaining interest, and workshops specializing on machine learning in health care typically boast more than 300 attendees from various fields. MICCAI offers an ideal and timely opportunity to combine this rising interest in data-driven health care with medical imaging expertise. Beyond MIC assembles researchers of different specializations and shared interests in this newly evolving field, facilitating the advancement of novel methods and technologies. Specifically, the mathematical, statistical, and algorithmic thinking, and image processing experience of the MICCAI community can help develop new methods for the analysis of emerging imaging and non-imaging modalities. Beyond MIC offers an ideal meeting to bridge the gap between the various communities that can contribute to these solutions. Beyond MIC includes keynote sessions introducing the state of the art and challenges of the field, as well as presentations of accepted papers published here, discussing novel methods or new applications.

August 2018

Adrian V. Dalca
Mert R. Sabuncu
Li Shen

Organization

Organizing Committee

Adrian V. Dalca   Mass. General Hospital, Harvard Medical School and
     CSAIL, Mass. Institute of Technology, USA

Mert R. Sabuncu   School of Electrical and Computer Engineering,
     and Nancy E. and Peter C. Meinig School of
     Biomedical Engineering, Cornell University, USA

Li Shen   The Perelman School of Medicine, University of
     Pennsylvania, USA

Contents

Proceedings of the Second Workshop on GRaphs in biomedicAl Image anaLysis

Graph Saliency Maps Through Spectral Convolutional Networks: Application to Sex Classification with Brain Connectivity

Salim Arslan[⊠], Sofia Ira Ktena, Ben Glocker, and Daniel Rueckert

Biomedical Image Analysis Group, Department of Computing,
Imperial College London, London, UK
s.arslan@imperial.ac.uk

Abstract. Graph convolutional networks (GCNs) allow to apply traditional convolution operations in non-Euclidean domains, where data are
commonly modelled as irregular graphs. Medical imaging and, in particular, neuroscience studies often rely on such graph representations,
with brain connectivity networks being a characteristic example, while
ultimately seeking the locus of phenotypic or disease-related differences
in the brain. These regions of interest (ROIs) are, then, considered to
be closely associated with function and/or behaviour. Driven by this, we
explore GCNs for the task of ROI identification and propose a visual
attribution method based on class activation mapping. By undertaking
a sex classification task as proof of concept, we show that this method
can be used to identify salient nodes (brain regions) without prior node
labels. Based on experiments conducted on neuroimaging data of more
than 5000 participants from UK Biobank, we demonstrate the robustness of the proposed method in highlighting reproducible regions across
individuals. We further evaluate the neurobiological relevance of the identified regions based on evidence from large-scale UK Biobank studies.

1 Introduction

Graph convolutional neural networks (GCNs) have recently gained a lot of attention, as they allow adapting traditional convolution operations from Euclidean
to irregular domains [1]. Irregular graphs are encountered very often in medical imaging and neuroscience studies in the form of brain connectivity networks,
supervoxels or meshes. In these cases, applications might entail both node-centric
tasks, e.g. node classification, as well as graph-centric tasks, e.g. graph classification or regression. While CNNs have redefined the state-of-the-art in numerous
problems by achieving top performance in diverse computer vision and pattern
recognition tasks, insights into their underlying decision mechanisms and the
impact of the latter on performance are still limited.

Recent works in deep learning address the problem of identifying salient
regions in 2D/3D images in order to visualise determinant patterns for classifi-
cation/regression tasks performed by a CNN and obtain spatial information that

© Springer Nature Switzerland AG 2018
D. Stoyanov et al. (Eds.): GRAIL 2018/Beyond MIC 2018, LNCS 11044, pp. 3–13, 2018.
https://doi.org/10.1007/978-3-030-00689-1_1

might be useful for the delineation of regions of interest (ROI) [2]. In the field of neuroscience, in particular, the identification of the exact locus of disease- or phenotype-related differences in the brain is commonly sought. Locating brain areas with a critical role in human behaviour and mapping functions to brain regions as well as diseases on disruptions to specific structural connections are among the most important goals in the study of the human connectome.

In this work, we explore GCNs for the task of brain ROI identification. As proof of concept, we undertake a sex classification task on functional connectivity networks, since there is previous evidence for sex-related differences in brain connectivity [3]. Characteristically, stronger functional connectivity was established within the default mode network of female brains, while stronger functional connectivity was found within the sensorimotor and visual cortices of male brains [4]. As a result, we consider this a suitable application to demonstrate the potential of the proposed method for delineating brain regions based on the attention/sensitivity of the model to the sex of the input subject's connectivity graph. More specifically, we show that spatially segregated salient regions can be identified in non-Euclidean space by using class activation mapping [5] on GCNs, making it possible to effectively map the most important brain regions for the task under consideration.

Related Work: Graph convolutions have been employed to address both graph-centric and node-centric problems and can be performed in the spatial [6] or spectral domain [7,8]. In the latter case, convolutions correspond to multiplications in the graph spectral domain and localised filters can be obtained with Chebyshev polynomials [7] or rational complex functions [8]. [9] introduced adaptive graph convolutions and attention mechanisms for graph- and node-centric tasks, while in [10] attention mechanisms were employed to assign different weights to neighbours in node classification tasks with inductive and transductive inference. Although the latter works focus the attention of the network onto the most relevant nodes, they overlook the importance/contribution of different features/graph elements for the task at hand.

At the same time, visual feature attribution through CNNs has attracted attention, as it allows identifying salient regions in an input image that lead a classification network to a certain prediction. It is typically addressed with gradient and/or activation-based strategies. The former relies on the gradients of the prediction with respect to the input and attributes saliency to the regions that have the highest impact on the output [2]. Activation-based methods, on the other hand, associate feature maps acquired in the final convolutional layer with particular classes and use weighted activations of the feature maps to identify salient regions [5]. A recent work addresses the problem from an adversarial point of view and proposes a visual attribution technique based on Wasserstein generative adversarial networks [11]. While these methods offer promising results on Euclidean images, their application to graph-structured data is yet to be explored.

Contributions: We propose a visual feature attribution method for graph-structured data by combining spectral convolutional networks and class activation mapping [5]. Through a graph classification task, in which each graph represents a brain connectivity network, we detect and visualise brain regions that are responsible for the prediction of the classifier, hence providing a new means of brain ROI identification. As a proof of concept, we derive experiments in the context of sex differences in functional connectivity. First, we train a spectral convolutional network classifier and achieve state-of-the-art accuracy in the prediction of female and male subjects based on their functional connectivity networks captured at rest. The activations of the feature maps are, then, used for visual attribution of the nodes, each of which is associated with a brain region. Using resting-state fMRI (rs-fMRI) data of more than 5000 subjects acquired by UK Biobank, we show that the proposed method is highly robust in selecting the same set of brain regions/nodes across subjects and yields highly reproducible results across multiple runs with different seeds.

2   Method

Figure 1 illustrates the proposed method for identifying brain regions used by GCNs to predict a subject's sex based on its functional connectivity. Given an adjacency matrix that encodes similarities between nodes and a feature matrix representing a node's connectivity profile, the proposed method outputs the sex of the input subject and provides a graph saliency map highlighting the brain regions/nodes that lead to the corresponding prediction. Finally, we rank brain regions with respect to their contribution towards driving the model's prediction at subject level and compute a population-level saliency map by combining them across individuals.

Spectral Graph Convolutions: We assume n samples (*i.e.* subjects), $X = [X_1, \ldots, X_n]^T$, with signals defined on a graph structure. Each subject is associated with a data matrix $X_i \in \mathbb{R}^{d_x \times d_y}$, where d_y is the dimensionality of the node's feature vector (*i.e.* signal), and a label $y_i \in \{0, 1\}$. In order to encode the structure of the data, we define a weighted graph $G = (V, E, W)$ where V is the set of $d_x = |V|$ nodes (vertices), E is the set of edges (connections) and $W \in \mathbb{R}^{d_x \times d_x}$ is the weighted adjacency matrix, representing the weight of each edge, *i.e.* $W_{i,j}$ is the weight of the edge connecting $v_i \in V$ to $v_j \in V$.

A convolution in the graph spatial domain corresponds to a multiplication in the graph spectral domain. Hence, graph filtering operations can be performed in the spectral domain using the eigenfunctions of the normalised Laplacian of a graph [12], which is defined as $L = I_{d_x} - D^{-\frac{1}{2}} W D^{-\frac{1}{2}}$, where D is the degree matrix and I_{d_x} the identity matrix. In order to yield filters that are strictly localised and efficiently computed, Defferrard et al. [7] suggested a polynomial parametrisation on the Laplacian matrix by means of Chebyshev polynomials. Chebyshev polynomials are recursively computed using $T_k(L) = 2LT_{k-1}(L) - T_{k-2}$, with $T_0(L) = 1$ and $T_1(L) = L$.

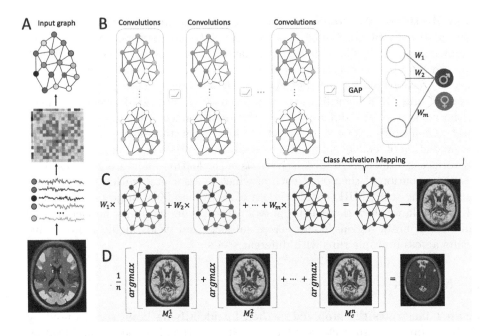

Fig. 1. Overview of the proposed approach. (A) The input graph is computed using a brain parcellation and rs-fMRI connectivity signals. (B) Graph convolutional network model. Convolutional feature maps of the last layer are spatially pooled via global average pooling (GAP) and connected to a linear sex classifier. (C) Class activation mapping procedure. (D) Generation of population-level saliency maps.

A polynomial of order K yields strictly K-localised filters. Filtering of a signal x with a K-localised filter can, then, be performed using:

$$y = g_\theta(L) * x = \sum_{k=0}^{K} \theta_k T_k(\tilde{L})x, \tag{1}$$

with $\tilde{L} = \frac{2}{\lambda_{max}}L - I_{d_x}$ and λ_{max} denoting the largest eigenvalue of the normalised Laplacian, L. The output of the lth layer for a sample s in a graph convolutional network is, then, given by:

$$y_s^l = \sum_{i=1}^{F_{in}} g_{\theta_i^l}(L)x_{s,i}^l. \tag{2}$$

For F_{out} output filter banks and F_{in} input filter banks, this yields $F_{in} \times F_{out}$ vectors of trainable Chebyshev coefficients $\theta_i^l \in \mathbb{R}^K$ with $x_{s,i}^l$ denoting the input feature map i for sample s at layer l. Hence, at each layer the total number of trainable parameters is $F_{in} \times F_{out} \times K$.

Class Activation Mapping: Class activation mapping (CAM) [5] is a technique used to identify salient regions that assist a CNN to predict a particular class. It builds on the fact that, even though no supervision is provided on the object locations, feature maps in various layers of CNNs still provide reliable localisation information [13], which can be captured via global average pooling (GAP) in the final convolutional layer [14]. Encoded into a class activation map, these "spatially-averaged" deep features not only yield approximate locations of objects, but also provide information about where the attention of the model focuses when predicting a particular class [5]. In the context of GCNs, CAM is used to localise discriminative nodes, each associated with a saliency score.

The process for generating class activation maps is illustrated in Fig. 1. Given a typical GCN model which consists of a series of convolutional layers, a GAP layer is inserted into the network right after the last convolutional layer. The spatially-pooled feature maps are connected to a dense layer that produces the output for a classification task (Fig. 1B). We can then linearly map the weights of the dense layer onto the corresponding feature maps to generate a class activation map showing the salient nodes in the graph (Fig. 1C).

More formally, let $f_i(v)$ represent the activation of the ith feature map in the last convolutional layer at node v. For the feature map i, the average pooling operation is defined as $F_i = (1/d_z) \sum_v f_i(v)$, where $F_i \in \mathbb{R}$ and d_z is the number of nodes in the feature map. Thus, for a given class c, the input to the dense layer is $\sum_i w_i^c F_i$, where w_i^c is the corresponding weight of F_i for class c. Intuitively, w_i^c indicates the importance of F_i for class c, therefore, we can use these weights to compute a class activation map M_c, where each node is represented by a weighted linear sum of activations, *i.e.* $M_c(v) = \sum_i w_i^c f_i(v)$. This map shows the impact of a node v to the prediction made by the GCN model and, once projected back onto the brain, can be used to identify the ROIs that are most relevant for the specific classification task.

Population-Level Saliency Maps: Although CAM provides graph-based activation maps at subject/class-level, population-level statistics about discriminative brain regions are also important. In order to combine class activation maps across subjects, we define a simple *argmax* operation that, for each subject, returns the index of the k top nodes with the highest activation. These are, subsequently, averaged across subjects and referenced as the population-level saliency maps as illustrated in Fig. 1D.

Network Architecture and Training: The details of the GCN architecture are presented in Table 1 and summarised as follows. 5 convolutional layers, each succeeded by rectified linear (ReLU) non-linearity are used. No pooling is performed between consecutive layers, as empirical results suggest that reducing the resolution of the underlying graph does not improve performance. We apply zero-padding to keep the spatial resolution of the feature maps unchanged throughout the model. A dropout rate of 0.5 is used in the 2nd, 4th, and 5th layers. The feature maps of the last layer are spatially averaged and connected

Table 1. Network architecture of the proposed model. * indicates the use of dropout for the corresponding convolutional layer.

Layer	Input	Conv	Conv*	Conv	Conv*	Conv*	GAP	Linear
Channels	55	32	32	64	64	128	128	2
K-order	N/A	9	9	9	9	9	N/A	N/A
Stride	N/A	1	1	1	1	1	N/A	N/A

to a linear classifier with softmax output. We employ global average pooling as it reflects where the attention of the network is focused and substantially reduces the number of parameters, hence alleviating over-fitting issues [14].

The loss function used to train the model comprises a cross entropy term and an L_2 regularisation term with decay rate of $5e^{-4}$. We use an Adam optimiser with momentum parameters $\beta = [0.9, 0.999]$ and initialise the training with a learning rate of 0.001. Training is performed for a fixed number of 500 steps (*i.e.* 20 epochs), in mini-batches of 200 samples, equally representing each class. We evaluate the model every 10 steps with an independent validation set, which is also used to monitor training. Based on this, the learning rate is decayed by a factor of 0.5, whenever validation accuracy drops in two consecutive evaluation rounds.

3   Data and Experiments

Dataset and Preprocessing: Imaging data is collected as part of UK Biobank's health imaging study (http://www.ukbiobank.ac.uk/), which aims to acquire imaging data for 100,000 predominantly healthy subjects. The multimodal scans together with the vast amount of non-imaging data are publicly available to assist researchers investigating a wide range of diseases, such as dementia, arthritis, cancer, and stroke. We conduct our experiments on rs-fMRI images available for 5430 subjects from the initial data release. Non-imaging data and medical information are also provided alongside brain scans including sex, age, genetic data, and many others. The dataset used here consists of 2873 female (aged 40–70 yo, mean 55.38 ± 7.41) and 2557 male (aged 40–70 yo, mean 56.61 ± 7.60) subjects.

Details of data acquisition and preprocessing procedures are given in [15]. Standard preprocessing steps have been applied to rs-fMRI images including motion correction, high-pass temporal filtering, and gradient distortion correction. An independent component analysis (ICA)-based approach is used to identify and remove structural artefacts [15]. Finally, images go through visual quality control and any preprocessing failures are eliminated.

Brain Parcellation and Network Modelling: A dimensionality reduction procedure known as "group-PCA" [16] is applied to the preprocessed data to obtain a group-average representation. This is fed to group-ICA [17] to parcellate the brain into 100 spatially independent, non-contiguous components. Group-ICA components identified as artefactual (*i.e.* not neuronally-driven) are discarded and the remaining $d = 55$ components are used to estimate a functional connectivity network for each subject by means of L_2-regularised partial correlation between the ICA components' representative timeseries [18] Each connectivity network corresponds to the data matrix X_i, i.e. $d_x = d_y$ in our application, and their average across training subjects is used to define the weighted graph W, in which only the $k = 10$ nearest neighbours are retained for each node, so that the local connectivity structure in the graph is effectively represented.

Experimental Setup: We use stratified 10-fold cross-validation to evaluate the model with split ratios set to 0.8, 0.1, and 0.1 for training, validation, and testing, respectively. Cross-validation allows to use all subjects for both training/validation and testing, while each subject in the dataset is used for testing exactly once. To further evaluate how the performance varies across different sets of subjects and how robust the identified salient regions are, we repeat cross-validation 10 times with different seeds.

4 Results and Discussion

In Table 2 we provide classification results obtained with the GCN classifier. The presented accuracy rates correspond to the results of all 10 folds for each run. On average, we achieve a test accuracy of 88.06% across all runs/folds, with low standard deviation for each run, indicating reproducible classification performance. While classification is not the end goal of the proposed method, a high accuracy rate is a prerequisite for robust and reliable activation maps. Yet, the average performance of our classifier is slightly higher than the state-of-the-art sex classification accuracy with respect to functional connectivity [19,20].

Figure 2 shows the sex-specific activations for all nodes. As illustrated, the GCN focuses on the same regions for both classes, with one class (*i.e.* female) consistently yielding higher activation than the other (*i.e.* male). This can be attributed to the fact that a binary classifier only needs to predict one class, while every other sample is automatically assigned the remaining class label. The most important nodes, in descending order, are 21, 5, 13, and 7. As indicated by the size of their markers, these four nodes are almost always ranked within the top $k = 3$ of all nodes with respect to their activations, meaning that all subjects

Table 2. Average sex classification accuracy rates (in %) for each run.

Run	1	2	3	4	5	6	7	8	9	10	Avr
Acc	88.51	88.27	87.64	87.94	87.84	88.01	88.51	88.08	88.05	87.77	**88.06**
Std	1.57	1.25	1.88	1.30	1.73	1.66	1.54	1.34	1.93	1.06	**1.57**

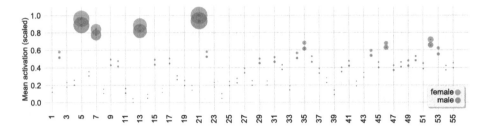

Fig. 2. Sex-specific class activations for all nodes averaged across subjects and runs. Mean activations are scaled to $[0, 1]$ for better visualisation. The size of the markers indicates the number of times a node is ranked within the top $k = 3$ most important, summed across subjects and runs.

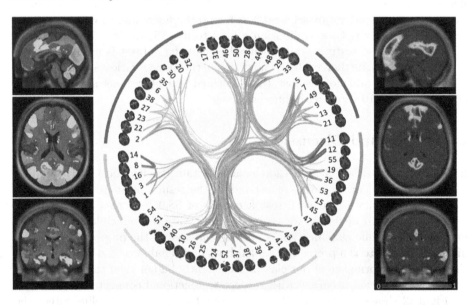

Fig. 3. Left: ICA-based brain parcellation shown in groups of six resting-state networks (RSNs), including the default mode network (red). The tree slices shown are, from top to bottom, sagittal, axial, and coronal, at indices 91, 112, and 91, respectively. Middle: Connectogram showing the group-averaged functional connectivity between 55 brain regions, which are clustered based on their average population connectivity. Strongest positive and negative correlations are shown in red and blue, respectively. Image is adapted from http://www.fmrib.ox.ac.uk/ukbiobank/netjs_d100/ and enhanced for better visualisation Right: Population-level saliency maps, combined for both genders. (Color figure online)

but few are consistently classified according to the connections of these nodes. While we only provide results for $k = 3$, the same regions are identified for lower/higher values of k, with only minimal changes in their occurrence rate, as shown in Supplementary Fig. 1.

In order to explore the neurobiological relevance of these results, we refer to the UK Biobank group-averaged functional connectome [15], which maps the functional interactions between the 55 brain regions clustered into six resting state networks (RSNs) according to their average population connectivity (Fig. 3). RSNs comprise spatially segregated, but functionally connected cortical regions, that are associated with diverse functions, such as sensory/motor, visual processing, auditory processing, and memory. Our comparisons to the connectome revealed that regions 21, 5, 13, and 7 (as shown in Fig. 3) are part of the default mode network (highlighted with red), a spatially distributed cluster which is activated 'by default' during rest. A large-scale study on sex differences in the human brain [4] has also found evidence that functional connectivity is stronger for females in the default mode network, which might further indicate that the identified regions are neurobiologically relevant and reflect sex-specific characteristics encoded in functional connectivity.

5   Conclusion

In this paper, we have addressed the visual attribution problem in graph-structured data and proposed an activation-based approach to identify salient graph nodes using spectral convolutional neural networks. By undertaking a graph-centric classification task, we showed that a GCN model enhanced with class activation mapping can be used to identify graph nodes (brain regions), even in the absence of supervision/labels at the node level. Based on experiments conducted on neuroimaging data from UK Biobank, we demonstrated the robustness of the proposed method by means of highlighting the same regions across different subjects/runs using cross validation. We further validated the neurobiological relevance of the identified ROIs based on evidence from UK Biobank studies [4,15].

While the potential of the proposed method is demonstrated on functional networks with rs-fMRI, it can be applied to any graph-structured data and/or modality. However, its applicability might be limited by several factors, including the definition and number of nodes (e.g. brain parcellation), network modelling, as well as node signal choices. It is also important to assess the robustness of the identified regions by disentangling the effect of the graph structure and the node features. While the method can successfully localise the salient regions, its lack of ability to visualise the most important features remains as a limitation compared to classical linear models. Future work will focus on the applicability of the method to other graph-centric problems (e.g. regression). For instance, a GCN model can be trained for age prediction and consequently used to identify brain regions for which connectivity is most affected with ageing. Another interesting direction entails extending this work for directed/dynamic, e.g. time-varying, graphs, as well as using it for biomarker identification.

Acknowledgements. This research has been conducted using the UK Biobank Resource under Application Number 12579 and funded by the EPSRC Doctoral Prize Fellowship funding scheme.

References

1. Bronstein, M.M., Bruna, J., LeCun, Y., Szlam, A., Vandergheynst, P.: Geometric deep learning: going beyond euclidean data. IEEE Sig. Process. Mag. **34**(4), 18–42 (2017)
2. Simonyan, K., Vedaldi, A., Zisserman, A.: Deep inside convolutional networks: visualising image classification models and saliency maps. arXiv preprint arXiv:1312.6034 (2013)
3. Satterthwaite, T.D., Wolf, D.H., et al.: Linked sex differences in cognition and functional connectivity in youth. Cereb. Cortex **25**(9), 2383–2394 (2014)
4. Ritchie, S.J., Cox, S.R., Shen, X., et al.: Sex differences in the adult human brain: evidence from 5,216 UK Biobank participants. bioRxiv (2017)
5. Zhou, B., Khosla, A., Lapedriza, A., Oliva, A., Torralba, A.: Learning deep features for discriminative localization. In: 2016 IEEE Conference on Computer Vision and Pattern Recognition (CVPR), pp. 2921–2929. IEEE (2016)
6. Monti, F., Boscaini, D., Masci, J., Rodola, E., Svoboda, J., Bronstein, M.M.: Geometric deep learning on graphs and manifolds using mixture model CNNs. In: Proceedings of CVPR, vol. 1, p. 3 (2017)
7. Defferrard, M., Bresson, X., Vandergheynst, P.: Convolutional neural networks on graphs with fast localized spectral filtering. In: Advances in Neural Information Processing Systems, pp. 3844–3852 (2016)
8. Levie, R., Monti, F., Bresson, X., Bronstein, M.M.: CayleyNets: graph convolutional neural networks with complex rational spectral filters. arXiv preprint arXiv:1705.07664 (2017)
9. Zhou, Z., Li, X.: Convolution on graph: a high-order and adaptive approach (2018)
10. Veličković, P., Cucurull, G., Casanova, A., Romero, A., Liò, P., Bengio, Y.: Graph attention networks. arXiv preprint arXiv:1710.10903 (2017)
11. Baumgartner, C.F., Koch, L.M., Tezcan, K.C., Ang, J.X., Konukoglu, E.: Visual feature attribution using Wasserstein GANs. arXiv preprint arXiv:1711.08998 (2017)
12. Shuman, D.I., Narang, S.K., Frossard, P., Ortega, A., Vandergheynst, P.: The emerging field of signal processing on graphs: extending high-dimensional data analysis to networks and other irregular domains. IEEE Sig. Process. Mag. **30**(3), 83–98 (2013)
13. Zhou, B., Khosla, A., Lapedriza, A., Oliva, A., Torralba, A.: Object detectors emerge in deep scene CNNs. arXiv preprint arXiv:1412.6856 (2014)
14. Lin, M., Chen, Q., Yan, S.: Network in network. arXiv preprint arXiv:1312.4400 (2013)
15. Miller, K.L., Alfaro-Almagro, F., Bangerter, N.K., et al.: Multimodal population brain imaging in the UK Biobank prospective epidemiological study. Nat. Neurosci. **19**(11), 1523 (2016)
16. Smith, S.M., Hyvärinen, A., Varoquaux, G., Miller, K.L., Beckmann, C.F.: Group-PCA for very large fMRI datasets. Neuroimage **101**, 738–749 (2014)
17. Beckmann, C.F., Smith, S.M.: Probabilistic independent component analysis for functional magnetic resonance imaging. IEEE TMI **23**(2), 137–152 (2004)

18. Smith, S.M., et al.: Network modelling methods for fMRI. Neuroimage **54**(2), 875–891 (2011)
19. Ktena, S.I., et al.: Metric learning with spectral graph convolutions on brain connectivity networks. NeuroImage **169**, 431–442 (2018)
20. Arslan, S., Ktena, S.I., Makropoulos, A., Robinson, E.C., Rueckert, D., Parisot, S.: Human brain mapping: a systematic comparison of parcellation methods for the human cerebral cortex. NeuroImage **170**, 5–30 (2018). Segmenting the Brain

A Graph Representation and Similarity Measure for Brain Networks with Nodal Features

Yusuf Osmanlıoğlu[1]([✉]), Birkan Tunç[2], Jacob A. Alappatt[1], Drew Parker[1],
Junghoon Kim[3], Ali Shokoufandeh[4], and Ragini Verma[1]

[1] Center for Biomedical Image Computing and Analytics, Department of Radiology,
University of Pennsylvania, Philadelphia, USA
yusuf.osmanlioglu@uphs.upenn.edu
[2] Center for Autism Research, Children's Hospital of Philadelphia, Philadelphia, USA
[3] CUNY School of Medicine, The City College of New York, New York, USA
[4] Department of Computer Science, Drexel University, Philadelphia, USA

Abstract. The human brain demonstrates a network structure that is commonly represented using graphs with pseudonym connectome. Traditionally, connectomes encode only inter-regional connectivity as edges, while regional information, such as centrality of a node that may be crucial to the analysis, is usually handled as statistical covariates. This results in an incomplete encoding of valuable information. In order to alleviate such problems, we propose an enriched connectome encoding regional properties of the brain network, such as structural node degree, strength, and centrality, as node features in addition to representing structural connectivity between regions as weighted edges. We further present an efficient graph matching algorithm, providing two measures to quantify similarity between enriched connectomes. We demonstrate the utility of our graph representation and similarity measures on classifying a traumatic brain injury dataset. Our results show that the enriched representation combining nodal features and structural connectivity information with the graph matching based similarity measures is able to differentiate the groups better than the traditional connectome representation.

Keywords: Annotated brain networks · Brain graphs
Multi-feature representation · Graph matching

1 Introduction

Connectomes can be described as a graph of organized regions and their connections that putatively have foundational roles in emerging functional and cognitive outcomes [1]. Hence, many analyses in cognition, learning, and brain diseases and disorders investigate the organization of the brain [2]. Graph theoretical approaches such as complex network analysis provide powerful tools to study structural and functional characteristics of the brain without losing its organizational features [3].

© Springer Nature Switzerland AG 2018
D. Stoyanov et al. (Eds.): GRAIL 2018/Beyond MIC 2018, LNCS 11044, pp. 14–23, 2018.
https://doi.org/10.1007/978-3-030-00689-1_2

In traditional connectomes, when representing the brain as a network, the nodes of the network correspond to the brain regions, and the edges between the nodes correspond to connections between those regions. In this approach, networks encode only inter-regional connectivity. The regional information such as degree, strength, or centrality that may be crucial to the analysis are usually treated as confounding factors or covariates. This hinders interpretations regarding regional changes due to, for instance, an underlying pathology. However, graph theory facilitates a principled methodology to combine regional characteristics (node features) with interactions between regions (edge features), by means of annotating nodes of the network [4]. Hence, the first contribution of this study is to provide a rich brain network representation, an enriched connectome, that enables inclusion of such nodal features when modeling brain connectivity.

Such a rich representation of brain organization including nodal features requires a new set of tools such as a similarity measure between these networks (graphs) which is essential for classification, clustering, or regression tasks [5,6]. As a second contribution, we propose a graph matching algorithm that provides a similarity measure between brain networks with nodal features. Among several approaches proposed in the literature to calculate graph similarity over brain data such as seeded graph matching [7] and graph embedding [8], graph edit distance (GED) is arguably the most effective method with promising results [9,10]. However, high running time complexity of GED requires use of approximation techniques such as Hungarian algorithm in [11] and hinders a detailed analysis of edge features [12]. We approach the graph matching problem as an instance of the *metric labeling problem* [13] and provide an efficient approximation algorithm using the primal-dual scheme [14] by extending our previous study [15]. Our graph matching method achieves two goals simultaneously: finding a mapping between brain regions of different graphs and computing a similarity score. The enriched connectome along with the graph-based similarity measure facilitates its use in classification of samples and we demonstrate its effective application on a traumatic brain injury (TBI) dataset. Results show that our enriched connectome along with the proposed matching algorithm provides better classification between the groups than the traditional connectivity based connectome representation.

2 Materials and Method

2.1 Dataset

Participants: We use a traumatic brain injury dataset consisting 39 patients (12 female) with moderate-to-severe TBI examined at 3 months post injury and 30 healthy controls (8 female). Age of patients are in $[18, 65]$ years with a mean of 35 years and standard deviation of 14.7 years, while the age of healthy controls are in $[20, 56]$ years with a mean and standard deviation of 34.7 and 9.9 years, respectively. Duration of post-traumatic amnesia of patients, which can be considered as a measure of trauma severity, has a mean of 26.7 days with a standard deviation of 21.2 days.

Data Acquisition and Preprocessing: For each subject, DTI data was acquired on a Siemens 3T TrioTim scanner with a 8 channel head coil (single shot spin echo sequence, TR/TE $= 6500/84$ ms, $b = 1000$ s/mm^2, 30 gradient directions). 86 region of interests from the Desikan atlas [16] were extracted to represent the nodes of the structural network. A mask was defined using voxels with an FA of at least 0.1 for each subject. Deterministic tractography was performed to generate and select 1 million streamlines, seeded randomly within the mask. Angle curvature threshold of 60 degrees, and a min and max length threshold of 5 mm and 400 mm were applied, resulting in an 86×86 adjacency matrix of weighted connectivity values, where each element represents the number of streamlines between regions.

2.2   Enriched Connectome

Given parcellation of the brain into 86 regions, we constructed a weighted undirected graph with 86 nodes corresponding to brain regions and weighted edges corresponding to the number of fibers connecting region pairs. We annotate each node with two set of features. First, we generated a 6 dimensional feature vector by calculating various graph theoretical measures for each node, namely degree, strength, betweenness centrality, local efficiency, participation coefficient, and local assortativity, using the Brain Connectivity Toolbox [17]. While calculating participation coefficient of nodes, we used association of structural regions with 7 functional systems as described in [18]. Second, we generated an 86 dimensional feature vector, representing the weighted connectivity of each node to the rest of the nodes in the graph, where we considered self edges to be nil. In summary, our graph representation, denoted *enriched connectome* hereby, incorporates graph theoretical measures of the connectome along with the weighted connectivity of the regions that are to be found in network representations. We normalized the values of each graph theory measure and the edge weights to $[0, 1]$ in order to make them comparable across subjects.

2.3   Graph Matching Based Similarity Measure

We propose taking a graph matching approach to define a similarity measure between two enriched connectomes, while providing a mapping between their nodes. We note that since the brains are parcellated into a common atlas in our setup, mapping between the regions are known a priori. However, we expect to get several mismatching nodes between dissimilar enriched connectomes due to differences in the connectivity of the network, making the similarity of graphs and the ratio of mismatching nodes effective measures for identifying patients from controls.

To this end, we evaluate the graph matching as a special case of the metric labeling problem [13]. Translating the metric labeling into the domain of brain graphs, the problem reads as follows: Given two enriched connectomes A and B, find the optimal one-to-one mapping $f : A \rightarrow B$ between their nodes while minimizing the following objective function:

$$\beta \sum_{a \in \mathcal{A}} c(a, f(a)) + (1 - \beta) \sum_{a, a' \in \mathcal{A}} w(a, a') \cdot d(f(a), f(a')). \tag{1}$$

The first summation term in (1) is regarded as the *assignment cost* with $c : \mathcal{A} \times \mathcal{B} \to \mathbb{R}$ that determines the cost of mapping a brain region $a \in \mathcal{A}$ to a region $b \in \mathcal{B}$, which we define as $||v1_a - v1_b||_2 + ||v2_a - v2_b||_2$ where $v1$ and $v2$ indicate the graph theoretical and edge weight based feature vectors described earlier, respectively. The second summation term stands for the *separation cost,* penalizing strongly connected brain regions $a, a' \in \mathcal{A}$ in getting mapped to loosely connected regions $b, b' \in \mathcal{B}$ with $w : \mathcal{A} \times \mathcal{A} \to \mathbb{R}$ indicating edge weights in $\mathcal{A}$ as a measure of connectivity strength and $d : \mathcal{B} \times \mathcal{B} \to \mathbb{R}$ indicating the distance between nodes of $\mathcal{B}$ as a measure of proximity between regions which is defined inversely proportional to the w in $\mathcal{B}$. Thus, the first half of the cost function encourages mapping each node of $\mathcal{A}$ to a node that resembles it most in $\mathcal{B}$ while the second half discourages two strongly connected regions in $\mathcal{A}$ getting mapped to two loosely connected regions in $\mathcal{B}$. The variable β in (1) is a balancing term to adjust the contribution of the assignment and separation costs to the objective function which takes values in $[0, 1]$. Once optimized, summation of the costs in (1) defines a similarity score between the two graphs.

In their seminal paper, Kleinberg and Tardos presented the following quadratic optimization formulation for the metric labeling problem which they showed to be computationally intractable to solve [13]:

$$\min \sum_{\substack{a \in \mathcal{A} \\ b \in \mathcal{B}}} c(a, b) \cdot x_{a,b} + \sum_{\substack{a, a' \in \mathcal{A} \\ b, b' \in \mathcal{B}}} w(a, a') \cdot d(b, b') \cdot x_{a,b} \cdot x_{a',b'}$$
$$\text{s.t.} \quad \sum_{b \in \mathcal{B}} x_{a,b} = 1, \qquad \forall a \in \mathcal{A}$$
$$\sum_{a \in \mathcal{A}} x_{a,b} = 1, \qquad \forall b \in \mathcal{B} \tag{2}$$
$$x_{a,b} \in \{0, 1\}, \qquad \forall a \in \mathcal{A}, b \in \mathcal{B}$$

where $x_{a,b}$ is an indicator variable taking value 1 if a is mapped to b and 0 otherwise. They also presented a linear programming formulation of the problem along with an approximation algorithm using hierarchically well-separated trees (HST), which was inefficient due to the computational time it takes to build the HSTs and to solve the linear program. Using another integer linear programming formulation of the problem along with a primal-dual approximation scheme [14], we recently presented an efficient approximation algorithm for the traditional metric labeling problem [15]. Here, we extend the latter study by altering the constraints of the metric labeling to account for the particular case of matching the enriched connectomes. Traditional metric labeling formulation allows many-to-one matching of the nodes between graphs, that is, several nodes of the first graph can be mapped to the same node in the second graph. In the setup of enriched connectomes where the brains are registered to a common atlas and parcellated into the same number of regions across subjects, it is known a priori that there should be a one-to-one mapping between the nodes of the graphs. Motivated by this observation, we impose additional constraints to the metric labeling formulation to enforce a one-to-one mapping between graphs.

Our extended version of the metric labeling with the integer linear programming formulation is as follows:

$$\min \sum_{\substack{a \in \mathcal{A} \\ b \in \mathcal{B}}} c(a,b) \cdot x_{a,b} + \sum_{\substack{a,a' \in \mathcal{A} \\ b,b' \in \mathcal{B}}} w(a,a') \cdot d(b,b') \cdot x_{a,b,a',b'}$$

$$\begin{aligned}
\text{s.t.} \quad & \sum_{b \in \mathcal{A}} x_{a,b} = 1, & \forall a \in \mathcal{A} \\
& \sum_{a \in \mathcal{B}} x_{b,a} = 1, & \forall b \in \mathcal{B} \\
& \sum_{a' \in \mathcal{A}} x_{a,b,a',b'} = x_{a,b}, & \forall a \in \mathcal{A}, b, b' \in \mathcal{B} \\
& \sum_{b' \in \mathcal{B}} x_{a,b,a',b'} = x_{a,b}, & \forall a, a' \in \mathcal{A}, b' \in \mathcal{B} \\
& x_{a,b,a',b'} = x_{a',b',a,b}, & \forall a \neq a' \in \mathcal{A}, b \neq b' \in \mathcal{B} \\
& x_{a,b} \in \{0,1\}, x_{a,b,a',b'} \in \{0,1\}, & \forall a, a' \in \mathcal{A}, b, b' \in \mathcal{B}.
\end{aligned} \tag{3}$$

Note that, the formulation (3) replaces the quadratic term $x_{a,b} \cdot x_{a',b'}$ in (2) with the indicator variable $x_{a,b,a',b'}$, introducing $O(n^4)$ new variables and $O(n^3 + n^4)$ additional constraints relative to the linear programming formulation. Despite the increase in the size of the problem, this formulation allows applying the primal-dual scheme to efficiently achieve approximate results.

In order to get a primal-dual approximation algorithm for solving the metric labeling in its extended version in (3), we first state the dual of the formulation as follows:

$$\max \sum_{a \in \mathcal{A}} y_a + \sum_{b \in \mathcal{B}} y_b$$

$$\begin{aligned}
\text{s.t.} \quad & y_a + y_b - \sum_{a' \in \mathcal{A}} y_{a,b,a'} - \sum_{b \in \mathcal{B}} y_{a,b,b'} \leq c_{a,b}, \, \forall a \in \mathcal{A}, b \in \mathcal{B} \\
& \left. \begin{array}{l} y_{a,b,a'} + y_{a,b,b'} + y_{a,b,a',b'} \leq w_{a,a'} \cdot d_{b,b'} \\ y_{a',b,a} + y_{a,b,b'} - y_{a,b,a',b'} \leq w_{a',a} \cdot d_{b,b'} \end{array} \right\}, \quad \forall a, a' \in \mathcal{A}, b, b' \in \mathcal{B} \\
& y_p, y_{a,b,a'}, y_{a,b,b'}, y_{a,b,a',b'} \text{ unrestricted}, & \forall a, a' \in \mathcal{A}, b, b' \in \mathcal{B} \\
& y_a \geq 0, y_b \geq 0, & \forall a \in \mathcal{A}, b \in \mathcal{B}
\end{aligned} \tag{4}$$

Since the variables of type $y_{a,b,a',b'}$ appears as a summation and a subtraction in the second type of constraints of (4) which accounts for the balancing constraints in (3), strictly following the primal-dual method presented in [14] would require making assignments in tuples since it enforces dual feasibility throughout the algorithm, resulting in poor performance. As we previously suggested in [15], we relax the dual feasibility condition for the first type of the dual constraints that previously became tight and present an efficient primal-dual approximation algorithm for the problem in Algorithm 1.

The algorithm starts by initializing indicative variables $x_{a,b}$, set of unassigned nodes $\hat{\mathcal{A}}$ and $\hat{\mathcal{B}}$, and an adjusted assignment cost function ϕ where the value of $\phi(a,b)$ is initially set to be the assignment cost of a to b (line 1). In each iteration of the loop in lines 2–7, the algorithm maps a node a to a node b that minimizes the adjusted assignment cost function ϕ (lines 3–4). Before proceeding to the next iteration, assigned nodes a and b are removed from the sets $\hat{\mathcal{A}}$ and $\hat{\mathcal{B}}$ (line 5) and ϕ function is updated for each of the unassigned nodes in the set $\hat{\mathcal{A}}$ by an amount of separation cost with respect to the recently assigned nodes (line 6). Algorithm iterates until no unassigned node is left in $\hat{\mathcal{A}}$.

Algorithm 1. A primal-dual approximation algorithm for approximating (3)

---
  procedure Graph-match($\mathcal{P}, \mathcal{L}$)
1: $\forall a, a' \in \mathcal{A}, b \in \mathcal{B} : x_{a,b} \leftarrow 0, \hat{\mathcal{A}} \leftarrow \mathcal{A}, \hat{\mathcal{B}} \leftarrow \mathcal{B}$
           $\phi(a, b) \leftarrow c_{a,b}$
2: **while** $\hat{\mathcal{A}} \neq \emptyset$ **do**
3:     Find $a \in \hat{\mathcal{A}}$ that minimizes $\phi(a, b)$ for some $b \in \hat{\mathcal{B}}$
4:     $x_{a,b} \leftarrow 1$
5:     $\hat{\mathcal{A}} \leftarrow \hat{\mathcal{A}} \setminus \{a\}, \hat{\mathcal{B}} \leftarrow \hat{\mathcal{B}} \setminus \{b\}$
6:     $\forall a' \in \hat{\mathcal{A}}, b' \in \hat{\mathcal{B}} : \phi(a', b') = \phi(a', b') + w_{a,a'} \cdot d_{b,b'}$
7: **end while**
8: **return** $\mathcal{X} = \{x_{a,b} : \forall a \in \mathcal{A}, b \in \mathcal{B}\}$

We note that, $\phi(a, b)$ is not updated for a node a once it gets assigned, rendering the summation $\sum_{a,b} \phi(a, b) x_{a,b}$ to be equal to the similarity score between the two graphs since it is equal to the value of the objective function in (4) which in turn is equal to the value of the objective function in (3).

3 Results

Here, we demonstrate the utility of our brain network representation and similarity measure on a TBI dataset, where the goal is the binary classification of subjects into healthy controls and TBI patients. We used k-nearest neighbors (kNN) classifier.

3.1 TBI Classification

We used nested leave-one-out approach for cross validation, due to limited number of subjects. For each subject in the dataset, we used the remaining 68 subjects of the dataset as the training set. Using an inner leave-one-out cross validation with training set, we decided the balancing parameter β and the number of neighbors k to be used in the nearest neighbor search. Then, we tested each subject with the learned parameter tuple that achieved best classification accuracy.

For comparison purposes, we performed the experiment using two scenarios. First (baseline), we used only a traditional connectome where we represented edge weights in a vector form without a graph representation. Similarity between subjects is calculated using Euclidean distance between these vectorized edge weights (denoted L_2-dist). Second, we use enriched connectome with Algorithm 1 (denoted Graph-match). Note that, Graph-match allows regions of the first graph to get mapped to any one of the regions in the second graph while L_2-dist inherently assumes an identity matching between the nodes of two graphs. The comparison of two scenarios, i.e., traditional connectome with L_2-dist vs. enriched connectome with Graph-match, facilitates subsequent analysis and interpretation on regional matches between brains, possibly providing insights into TBI-induced regional differences.

Table 1. Classification results from leave-one-out cross validation for the two scenarios: traditional connectome with L_2-dist vs. enriched connectome with Algorithm 1.

Scenario	Accuracy	Sensitivity	Specificity
Traditional connectome & L_2-dist	66.7	51.28	86.67
Enriched connectome & Graph-match	72.46	71.19	73.33

Classification performance is presented in Table 1 for the two scenarios, showing overall accuracy, specificity, and sensitivity. Comparing overall accuracy of the two scenarios suggests that our graph representation with the similarity measure captures more information to decide about the classification than the baseline. As suggested from the results, incorporating nodal features into the representation along with connectivity information improves the classification accuracy. In addition to this, relaxing node mappings between enriched connectomes in Graph-match makes it possible to capture subtle regional alterations, possibly associated with injury, which is reflected by the increased classification performance of Graph-match. We also note that, our approach achieves similar performance for classifying patients and controls as the sensitivity and specificity have similar values whereas traditional connectome with L_2-dist performs poorly for classifying patients. The comparison of ROC curves presented in Fig. 1(a) demonstrates the improved performance of our method over the baseline.

Nested leave-one-out cross validation scheme results in 69 different parameter tuples (β, k) for our method and 69 k values for the baseline approach. In our experiments, we observed that parameter values were mostly consistent for our method across runs. Specifically, we observed that the inner loop of the experiment has chosen $\beta = 0.9$ without any exception and $k = 15$ with only five outliers out of 69 iterations for our method. This can be contrasted to $k = 6$ being chosen for the baseline approach along with 9 outliers, suggesting the robustness of our graph matching algorithm.

3.2   Effect of Feature Types

In order to demonstrate the contribution of graph theory measures and edge weights as node features, in Table 2, classification results for the brain networks with only graph theoretical features and only edge weights as features are contrasted to both feature types being combined in a single brain graph. We observe

Table 2. Classification results of brain graphs with only graph theoretical features and with only edge weights as features, using Graph-match as the similarity measure.

Node features	Accuracy	Sensitivity	Specificity
Graph theoretical measures alone	42.03	56.41	23.34
Edge weights alone	62.32	46.15	83.34

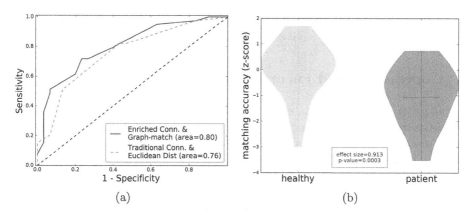

Fig. 1. (a) Comparison of ROC curves showing the classification performance of the baseline and the proposed method. (b) Z-score distribution of the matching accuracy for controls and patients with respect to the control population.

that combining both feature types improve the classification accuracy by 10% indicating that enriched connectome maintains more information by combining various features into a single structure relative to a network having either one of them as its only nodal feature. We also observe that using edge weights alone performs better than using graph theoretical measures alone, which can be attributed to larger number of features present in the former, providing a better feature set for classification. However, combination of the two providing an improvement over both of their individual classification accuracies indicate that the two sets of features represent unique aspects of the connectomics.

3.3   Mapping Between Nodes of Graphs

We note that, graph matching provides a mapping between nodes of the two graphs in addition to the similarity score between them. One might expect the regions of a brain graph to match their counterparts in another brain graph (such as, precentral gyrus in one enriched connectome would be expected to match with the precentral gyrus of another enriched connectome) as the brain anatomy is similar across people, with occasional mismatches due to subject-specific differences in connectivity. Leveraging this observation, we define another similarity measure, denoted *matching accuracy*, as the ratio of regions that are accurately matched with their counterparts to total number of regions. Matching enriched connectome of every subject to the healthy controls, we hypothesize that the matching accuracy of healthy controls with respect to themselves should be higher than the matching accuracy of patients with respect to healthy control population, as structural alterations introduced by TBI is expected to cause mismatching regions. As shown in Fig. 1(b), we observe a statistically significant group difference between the patients and controls in their matching accuracy with respect to the healthy subjects. We also observe that the matching accuracy is lower and has a larger variance in patient population, which can be attributed to altered structural connectivity due to pathology.

4   Conclusions and Future Work

In this paper, we presented an enriched connectome that allows combining multiple features into a single structure. The nodes in our representation correspond to the brain regions that are annotated with graph theoretical measures and connectivity of nodes with other nodes as node features, while the edges correspond to the structural connectivity between regions. We also proposed an efficient graph matching algorithm providing two similarity measures over our new representation, one being a summary measure of overall graph similarity and the other quantifying the ratio of number of accurately matched regions to total number of regions. Using the first measure, we showed that proposed enriched representation provided a better classification than the traditional connectomes, demonstrating contribution of the nodal features to information about the samples. Using the second measure, we demonstrated a significantly lower matching accuracy across patients relative to controls, highlighting trauma induced structural alterations in brains of patients.

In this study, we utilized features obtained from a single modality, namely DTI. Our graph representation can easily be extended to combine multiple modalities (e.g., DTI and fMRI). Adding multiple modalities introduce not only new nodal features, but also new edge types that will provide even a richer representation of the brain organization. Although the data that we used in this study involves known correspondences between connectomes, our method can also be applied on connectomes with unknown correspondences, as in subject specific parcellations.

Acknowledgements. This work was funded by NIH grants R01HD089390-01A1, 1 R01 NS096606, 5R01NS092398, and 5R01NS065980.

References

1. Tunç, B., Verma, R.: Unifying inference of meso-scale structures in networks. PLoS One **10**(11), e0143133 (2015)
2. Newman, M.E.J., Clauset, A.: Structure and inference in annotated networks. Nat. Commun. **7**, 11863 (2016)
3. Bullmore, E.T., Sporns, O., Solla, S.A.: Complex brain networks: graph theoretical analysis of structural and functional systems. Nat. Rev. Neurosci. **10**(3), 186–198 (2009)
4. Petersen, S.E., Sporns, O.: Brain networks and cognitive architectures. Neuron **88**(1), 207–219 (2015)
5. Van Den Heuvel, M.P., Pol, H.E.H.: Exploring the brain network: a review on resting-state fMRI functional connectivity. Eur. Neuropsychopharmacol. **20**(8), 519–534 (2010)

6. Chen, L., Vogelstein, J.T., Lyzinski, V., Priebe, C.E.: A joint graph inference case study: the *C. elegans* chemical and electrical connectomes. In: Worm, vol. 5, p. e1142041. Taylor & Francis (2016)
7. Fishkind, D.E., Adali, S., Priebe, C.E.: Seeded Graph Matching. arXiv preprint, arXiv:1209.0367v1 (2012)
8. Richiardi, J., Eryilmaz, H., Schwartz, S., Vuilleumier, P., Van De Ville, D.: Decoding brain states from fMRI connectivity graphs. Neuroimage 56(2), 616–626 (2011)
9. Ktena, S.I., Parisot, S., Passerat-Palmbach, J.: Comparison of Brain Networks with Unknown Correspondences. arXiv preprint arXiv:1611.04783, October 2016
10. Raj, A., Mueller, S.G., Young, K., Laxer, K.D., Weiner, M.: Network-level analysis of cortical thickness of the epileptic brain. Neuroimage 52(4), 1302–1313 (2010)
11. Riesen, K., Neuhaus, M., Bunke, H.: Bipartite graph matching for computing the edit distance of graphs. In: Escolano, F., Vento, M. (eds.) GbRPR 2007. LNCS, vol. 4538, pp. 1–12. Springer, Heidelberg (2007). https://doi.org/10.1007/978-3-540-72903-7_1
12. Abu-Aisheh, Z., Raveaux, R., Ramel, J.-Y., Martineau, P.: A parallel graph edit distance algorithm. Expert Syst. Appl. 94, 41–57 (2018)
13. Kleinberg, J., Tardos, É.: Approximation algorithms for classification problems with pairwise relationships: metric labeling and markov random fields. J. ACM 49(5), 616–639 (2002)
14. Goemans, M.X., Williamson, D.P.: The primal-dual method for approximation algorithms and its application to network design problems. In: Hochbaum, D.S. (ed.) Approximation Algorithms for NP-hard Problems, pp. 144–191. PWS Publishing Co., Boston (1997)
15. Osmanlıoğlu, Y., Ontañón, S., Hershberg, U., Shokoufandeh, A.: Efficient approximation of labeling problems with applications to immune repertoire analysis. In: 2016 23rd International Conference on Pattern Recognition (ICPR), pp. 2410–2415. IEEE (2016)
16. Desikan, R.S., et al.: An automated labeling system for subdividing the human cerebral cortex on MRI scans into gyral based regions of interest. Neuroimage 31(3), 968–980 (2006)
17. Rubinov, M., Sporns, O.: Complex network measures of brain connectivity: uses and interpretations. Neuroimage 52(3), 1059–1069 (2010)
18. Yeo, B.T.T., et al.: The organization of the human cerebral cortex estimated by intrinsic functional connectivity. J. Neurophysiol. 106(3), 1125–1165 (2011)

Multi-modal Disease Classification in Incomplete Datasets Using Geometric Matrix Completion

Gerome Vivar[1,2]([⊠]), Andreas Zwergal[2], Nassir Navab[1],
and Seyed-Ahmad Ahmadi[2]

[1] Technical University of Munich (TUM), Munich, Germany
g.vivar@tum.de
[2] German Center for Vertigo and Balance Disorders (DSGZ),
Ludwig-Maximilians-Universität (LMU), Munich, Germany

Abstract. In large population-based studies and in clinical routine, tasks like disease diagnosis and progression prediction are inherently based on a rich set of multi-modal data, including imaging and other sensor data, clinical scores, phenotypes, labels and demographics. However, missing features, rater bias and inaccurate measurements are typical ailments of real-life medical datasets. Recently, it has been shown that deep learning with graph convolution neural networks (GCN) can outperform traditional machine learning in disease classification, but missing features remain an open problem. In this work, we follow up on the idea of modeling multi-modal disease classification as a matrix completion problem, with simultaneous classification and non-linear imputation of features. Compared to methods before, we arrange subjects in a graph-structure and solve classification through geometric matrix completion, which simulates a heat diffusion process that is learned and solved with a recurrent neural network. We demonstrate the potential of this method on the ADNI-based TADPOLE dataset and on the task of predicting the transition from MCI to Alzheimer's disease. With an AUC of 0.950 and classification accuracy of 87%, our approach outperforms standard linear and non-linear classifiers, as well as several state-of-the-art results in related literature, including a recently proposed GCN-based approach.

1 Introduction

In clinical practice and research, the analysis and diagnosis of complex phenotypes or disorders along with differentiation of their aetiologies rarely relies on a single clinical score or data modality, but instead requires input from various modalities and data sources. This is reflected in large datasets from well-known multi-centric population studies like the Alzheimer's Disease Neuroimaging Initiative (ADNI) and its derived TADPOLE grand challenge[1]. TADPOLE data, for example, comprises demographics, neuropsychological scores, functional and

---

[1] http://adni.loni.usc.edu || https://tadpole.grand-challenge.org/.

© Springer Nature Switzerland AG 2018
D. Stoyanov et al. (Eds.): GRAIL 2018/Beyond MIC 2018, LNCS 11044, pp. 24–31, 2018.
https://doi.org/10.1007/978-3-030-00689-1_3

morphological features derived from MRI, PET and DTI imaging, genetics, as well as histochemical analysis of cerebro-spinal fluid. The size and richness of such datasets makes human interpretation difficult, but it makes them highly suited for computer-aided diagnosis (CAD) approaches, which are often based on machine learning (ML) techniques [10,11,16]. Challenging properties for machine learning include e.g. subjective, inaccurate or noisy measurements or a high number of features. Linear [11] and non-linear [16] classifiers for CAD show reasonable success in compensating for such inaccuracies, e.g. when predicting conversion from mild-cognitive-impairment (MCI) to Alzheimer's disease (AD). Recent work has further shown that an arrangement of patients in a graph structure based on demographic similarity [12] can leverage network effects in the cohort and increase robustness and accuracy of the classification. This is especially valid when combined with novel methods from geometric deep learning [1], in particular spectral graph convolutions [7]. Similar to recent successes of deep learning methods in medical image analysis [8], deep learning on graphs shows promise for CAD, by modeling connectivity across subjects or features.

Next to noise, a particular problem of real-life, multi-modal clinical datasets is missing features, e.g. due to restrictions in examination cost, time or patient compliance. Most ML algorithms, including the above-mentioned, require feature-completeness, which is difficult to address in a principled manner [4]. One interesting alternative to address missing features is to model CAD and disease classification as a matrix completion problem instead. Matrix completion was proposed in [5] for simultaneously solving the three tasks of multi-label learning, transductive learning, and feature imputation. Recently, this concept was applied for CAD in multi-modal medical datasets for the first time [15], for prediction of MCI-to-AD conversion on ADNI data. The method introduced a pre-computed feature weighting term and outperformed linear classifiers on their dataset, however it did not yet leverage any graph-modeled network effects of the population as in [12]. To this end, several recent works incorporated a geometric graph structure into the matrix completion problem [6,9,13]. All these methods were applied on non-medical datasets, e.g. for recommender systems [9]. Hence, their goal was solely imputation, without classification. Here, we unify previous ideas in a single stream-lined method that can be trained end-to-end.

Contribution. In this work, we follow up on the idea of modeling multi-modal CAD as a matrix completion problem [5] with simultaneous imputation and classification [15]. We leverage cohort network effects by integrating a population graph with a solution based on geometric deep learning and recurrent neural networks [9]. For the first time, we demonstrate geometric matrix completion (GMC) and disease classification from multi-modal medical data, towards MCI-to-AD prediction from TADPOLE features at baseline examination. In this difficult task, GMC significantly outperforms regular linear and non-linear machine learning methods as well as three state-of-the-art results from related works, including a recent approach based on graph-convolutional neural networks.

2 Methods

2.1 Dataset and Preprocessing

As an example application, we utilize the ADNI-based TADPOLE dataset, with the goal of predicting whether an MCI subject will convert to AD given their baseline information. We select all unique subjects with baseline measurements from ADNI1, ADNIGO, and ADNI2 in the TADPOLE dataset which were diagnosed as MCI including those diagnosed as EMCI and LMCI. Following [15], we retrospectively label those subjects whose condition progressed to AD within 48 months as cMCI and those whose condition remained stable as sMCI. The remaining MCI subjects who progressed to AD after month 48 are excluded from this study. We use multi-modal features from MRI, PET, DTI, and CSF at baseline, i.e. excluding longitudinal features. We use all numerical features from this dataset to stack with the labels and include age and gender to build the graph, following the intuition and methodology from [12].

2.2 Matrix Completion

We will start by describing the matrix completion problem. Suppose there exists a matrix $\mathbf{Y} \in \mathbb{R}^{m \times n}$ where the values in this matrix are not all known. The goal is to recover the missing values in this matrix. A well-defined description of this problem is to assume that the matrix is of low rank [2],

$$\min_{\mathbf{X} \in \mathbb{R}^{m \times n}} \quad \text{rank}(\mathbf{X}) \text{ s.t. } x_{ij} = y_{ij}, \forall ij \in \Omega, \tag{1}$$

where $\mathbf{X}$ is the $m \times n$ matrix with values x_{ij}, Ω is the set of known entries in matrix $\mathbf{Y}$ with y_{ij} values. However, this rank minimization problem (1) is known to be computationally intractable. So instead of solving for rank($\mathbf{X}$), we can replace it with its convex surrogate known as the nuclear norm $||\mathbf{X}||_*$ which is equal to the sum of its singular values [2]. In addition, if the observations in Y have noise, the equality constraint in Eq. (1) can be replaced with the squared Frobenius norm $||.||_F^2$ [3],

$$\min_{\mathbf{X} \in \mathbb{R}^{m \times n}} \quad ||\mathbf{X}||_* + \frac{\gamma}{2}||\mathbf{\Omega} \circ (\mathbf{Y} - \mathbf{X})||_F^2, \tag{2}$$

where $\mathbf{\Omega}$ is the masking matrix of known entries in $\mathbf{Y}$ and $\circ$ is the Hadamard product. Alternatively, a factorized solution to the representation of the matrix $\mathbf{X}$ was also introduced in [13,14], as the formulation using the full matrix makes it hard to scale up to large matrices such as the famous Netflix challenge. Here, the matrix $\mathbf{X} \in \mathbb{R}^{m \times n}$ is factorized into 2 matrices $\mathbf{W}$ and $\mathbf{H}$ via SVD, where $\mathbf{W}$ is $m \times r$ and $\mathbf{H}$ is $n \times r$, with $r \ll \min(m, n)$. Srebro et al. [14] showed that the nuclear norm minimization problem can then be rewritten as:

$$\min_{W,H} \frac{1}{2}||\mathbf{W}||_F^2 + \frac{1}{2}||\mathbf{H}||_F^2 + \frac{\gamma}{2}||\mathbf{\Omega} \circ (\mathbf{W}\mathbf{H}^T - \mathbf{Y})||_F^2 \tag{3}$$

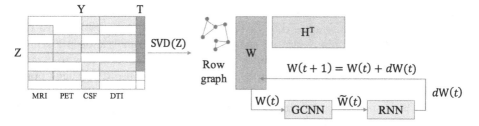

Fig. 1. Illustration of the overall approach: the matrix Z comprising incomplete features and labels is factorized into $Z = WH^T$. A connectivity graph is defined over rows W. During optimization, GCNN filters are learned along with RNN parameters and weight updates for W, towards optimal matrix completion of Z and simultaneous inference of missing features and labels in the dataset.

2.3 Matrix Completion on Graphs

The previous matrix completion problem can be extended to graphs [6,13]. Given a matrix $\mathbf{Y}$, we can assume that the rows/columns of this matrix are on the vertices of the graph [6]. This additional information can then be included into the matrix completion formulation in Eq. (2) as a regularization term [6]. To construct the graph, we can use meta-information out of these rows/columns or use the row/column vectors of this matrix to calculate a similarity metric between pairs of vertices. Given that every row in the matrix has this meta-information, Kalofolias et al. [6] showed that we can build an undirected weighted row graph $G_r = (V_r, E_r, A_r)$, with vertices $V_r = \{1, \ldots, m\}$. Edges $E_r \subseteq V_r \times V_r$ are weighted with non-negative weights represented by an adjacency matrix $A_r \in \mathbb{R}^{m \times m}$. The column graph $G_c = (V_c, E_c, A_c)$ is built the same way as the row graph, where the columns are now the vertices in G_c. Kalofolias et al. [6] showed that the solution to this problem is equivalent to adding the Dirichlet norms, $||\mathbf{X}||_{D,r}^2 = \text{tr}(X^T L_r X)$ and $||\mathbf{X}||_{D,c}^2 = \text{tr}(X L_c X^T)$, where L_r and L_c are the unnormalized row and column graph Laplacian, to Eq. (2),

$$\min_{\mathbf{X} \in \mathbb{R}^{m \times n}} \quad ||\mathbf{X}||_* \; + \; \frac{\gamma}{2}||\mathbf{\Omega} \circ (\mathbf{Y} - \mathbf{X})||_F^2 \; + \; \frac{\alpha_r}{2}||\mathbf{X}||_{D,r}^2 + \; \frac{\alpha_c}{2}||\mathbf{X}||_{D,c}^2 \quad (4)$$

The factorized formulation [9,13] of Eq. (4) is

$$\min_{\mathbf{W},\mathbf{H}} \frac{1}{2}||\mathbf{W}||_{D,r}^2 + \frac{1}{2}||\mathbf{H}||_{D,c}^2 + \frac{\gamma}{2}||\mathbf{\Omega} \circ (\mathbf{Y} - \mathbf{WH}^T)||_F^2 \quad (5)$$

2.4 Geometric Matrix Completion with Separable Recurrent Graph Neural Networks

In [9], Monti et al. propose to solve the matrix completion problem as a learnable diffusion process using Graph Convolutional Neural Networks (GCNN) and Recurrent Neural Networks (RNN). The main idea here is to use GCNN to extract features from the matrix and then use LSTMs to learn the diffusion process. They argue that combining these two methods allows the network to predict accurate small changes $\mathbf{dX}$ (or $\mathbf{dW}$, $\mathbf{dH}$ of the matrices $\mathbf{W}$, $\mathbf{H}$) to the matrix $\mathbf{X}$. Further details regarding the main ideas in geometric deep learning have been summarized in a review paper [1], where they elaborate how to extend convolutional neural networks to graphs. Following [9], we use Chebyshev polynomial basis on the factorized form of the matrix $\mathbf{X} = \mathbf{WH}^{\mathrm{T}}$ to represent the filters on the respective graph to each matrix $\mathbf{W}$ and $\mathbf{H}$. In this work, we only apply GCNN to the matrix $\mathbf{W}$ as we only have a row graph and leave the matrix $\mathbf{H}$ as a changeable variable. Figure 1 illustrates the overall approach.

2.5 Geometric Matrix Completion for Heterogeneous Matrix Entries

In this work, we propose to solve multi-modal disease classification as a geometric matrix completion problem. We use a Separable Recurrent GCNN (sRGCNN) [9] to simultaneously predict the disease and impute missing features on a dataset which has partially observed features and labels. Following Goldberg et al. [5], we stack a feature matrix $\mathbf{Y} \in \mathbb{R}^{m \times n}$ and a label matrix $\mathbf{T} \in \mathbb{R}^{m \times c}$ as a matrix $\mathbf{Z} \in \mathbb{R}^{m \times n+c}$, where m is the number of subjects, n is the dimension of the feature matrix, and c is the dimension of the target values. In the TADPOLE dataset, we stack the $m \times n$ feature matrix to the $m \times 1$ label matrix, where the feature matrix contains all the numerical features and the label matrix contains the encoded binary class labels for cMCI and sMCI. We build the graph by using meta-information from the patients such as their age and gender, similar to [12], as these information are known to be risk factors for AD. We compare two row graph construction approaches, first from age and gender information using a similarity metric [12] and second from age information only, using Euclidean distance-based k-nearest neighbors. Every node in a graph corresponds to a row in the matrix $\mathbf{W}$, and the row values to its associated feature vector. Since we only have a row graph, we leave the matrix $\mathbf{H}$ to be updated during backpropagation. To run the geometric matrix completion method we use the loss:

$$\ell(\Theta) = \frac{\gamma_a}{2}||\mathbf{W}||_{\mathrm{D},r}^2 + \frac{\gamma_b}{2}||\mathbf{W}||_{\mathrm{F}}^2 + \frac{\gamma_c}{2}||\mathbf{H}||_{\mathrm{F}}^2 + \frac{\gamma_d}{2}||\mathbf{\Omega}_a \circ (\mathbf{Z} - \mathbf{WH}^T)||_{\mathrm{F}}^2 + \gamma_e(\ell_{\mathbf{\Omega}_b}(\mathbf{Z}, \mathbf{X})),$$
(6)

where Θ are the learnable parameters, where $\mathbf{Z}$ denotes the target matrix, $\mathbf{X}$ is the approximated matrix, $||.||_{D,r}^2$ denotes the Dirichlet norm on a normalized row graph Laplacian, $\mathbf{\Omega}_a$ denotes the masking on numerical features, $\mathbf{\Omega}_b$ is the masking on the classification labels, and ℓ is the binary cross-entropy.

3   Results

We evaluate our approach on multi-modal TADPOLE data (MRI, PET, CSF, DTI) to predict MCI-to-AD conversion and compare it to several other multi-modal methods as baseline. We use a stratified 10-fold cross-validation strategy for all methods. Hyperparameters were optimized using Hyperopt[2], through

---

[2] http://hyperopt.github.io/hyperopt/.

nested cross-validation, targeting classification loss (binary cross-entropy) on a hold-out validation set (10% in each fold of training data). Following [9], we use the same sRGCNN architecture with parameters: rank $= 156$, chebyshev polynomial order $= 18$, learning rate $= 0.00089$, hidden-units $= 36$, $\gamma_a = 563.39$, $\gamma_b = 248.91$, $\gamma_c = 688.85$, $\gamma_d = 97.63$, and $\gamma_e = 890.14$.

It is noteworthy that at baseline, the data matrix Y with above-mentioned features is already feature-incomplete, i.e. only 53% filled. We additionally reduce the amount of available data randomly to 40%, 30% etc. to 5%. Figure 2 shows a comprehensive summary of our classification results in terms of area-under-the-curve (AUC). Methods we compare include mean imputation with random forest (RF), linear SVM (SVC) and multi-layer-perceptron (MLP), as well as three reference methods from literature [10,12,15], which operated on slightly different selections of ADNI subjects and on all available multi-modal features. While implementations of [10,15] are not publicly available, we tried to re-evaluate the method [12] using their published code. Unfortunately, despite our best efforts and hyperparameter optimization on our selection of TADPOLE data, we were not able to reproduce any AUC value close to their published value. To avoid any mistake on our side, we provide the reported AUC results rather than the worse results from our own experiments.

At baseline, our best-performing method with a graph setup based on age and gender ("GMC age-gender") [12] achieves classification with an AUC value of 0.950, compared to 0.902 [10], ~0.87 [12] and 0.851 [15]. In terms of classification accuracy, we achieved a value of 87%, compared to 82% [10] and 77% [12] (accuracy not reported in [15]). Furthermore, our method significantly outperforms standard classifiers RF, MLP and SVC at all levels of matrix completeness. The second graph configuration for our method ("GMC age" only) performs significantly worse and less stable than ("GMC age-gender"), confirming the usefulness

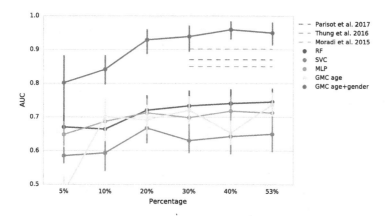

Fig. 2. Classification results: area under the curve (AUC) of our method, for different amounts of feature-completeness and in comparison to linear/non-linear standard methods, and three state-of-the-art results in literature (Parisot et al. [12], Thung et al. [15], Moradi et al. [10]).

of the row graph construction based on the subject-to-subject similarity measure proposed in [12]. Due to lower complexity of the GMC approach [9], training a single fold on recent hardware (Tensorflow on Nvidia GTX 1080 Ti) is on average 2x faster (11.8 s) than GCN (25.9 s) [12].

4   Discussion and Conclusion

In this paper, we proposed to view disease classification in multi-modal but incomplete clinical datasets as a geometric matrix completion problem. As an exemplary dataset and classification problem, we chose MCI-to-AD prediction. Our initial results using this method show that GMC outperforms three competitive results from recent literature in terms of AUC and accuracy. At all levels of additional random dropout of features, GMC also outperforms standard imputation and classifiers (linear and non-linear). There are several limitations which are worthy to be addressed. Results in Fig. 2 demonstrate that GMC is still sensitive to increasing amounts of feature incompleteness, in particular at feature presence below 15%. This may be due to our primary objective of disease classification during hyper-parameter optimization. For the same reason, we did not evaluate the actual imputation performed by GMC. However, an evaluation in terms of RMSE and a comparison to principled imputation methods [4] would be highly interesting, if this loss is somehow incorporated during hyperparameter optimization. Furthermore, we only evaluated GMC on ADNI data as represented in the TADPOLE challenge, due to the availability of multiple reference AUC/accuracy values in literature. As mentioned, however, disease classification in high-dimensional but incomplete datasets with multiple modalities is an abundant problem in computer-aided medical diagnosis. In this light, we believe that the promising results obtained through GMC in this study are of high interest to the community.

Acknowledgments. The study was supported by the German Federal Ministry of Education and Health (BMBF) in connection with the foundation of the German Center for Vertigo and Balance Disorders (DSGZ) (grant number 01 EO 0901).

References

1. Bronstein, M.M., Bruna, J., Lecun, Y., Szlam, A., Vandergheynst, P.: Geometric deep learning: going beyond euclidean data. IEEE Sig. Process. Mag. **34**(4), 18–42 (2017)
2. Candes, E.J., Recht, B.: Exact low-rank matrix completion via convex optimization. In: 46th Annual Allerton Conference on Communication, Control, and Computing, pp 1–49 (2008)
3. Candes, E.J., Plan, Y.: Matrix completion with noise. Proc. IEEE **98**(6), 925–936 (2010)
4. Dong, Y., Peng, C.Y.: Principled missing data methods for researchers. Springerplus **2**(1), 222 (2013)

5. Goldberg, A., Recht, B., Xu, J., Nowak, R., Zhu, X.: Transduction with matrix completion: three birds with one stone. In: Advances in Neural Information Processing Systems (NIPS), pp. 757–765 (2010)

6. Kalofolias, V., Bresson, X., Bronstein, M., Vandergheynst, P.: Matrix completion on graphs. arXiv:1408.1717 (2014)

7. Kipf, T.N., Welling, M.: Semi-supervised classification with graph convolutional networks. CoRR, arXiv:1609.02907 (2016)

8. Litjens, G., et al.: A survey on deep learning in medical image analysis. Med. Image Anal. **42**, 60–88 (2017)

9. Monti, F., Bronstein, M.M., Bresson, X.: Geometric matrix completion with recurrent multi-graph neural networks. CoRR, arXiv:1704.06803 (2017)

10. Moradi, E., Pepe, A., Gaser, C., Huttunen, H., Tohka, J.: Machine learning framework for early MRI-based Alzheimer's conversion prediction in MCI subjects. NeuroImage **104**, 398–412 (2015)

11. Oishi, K.: Multi-modal MRI analysis with disease-specific spatial filtering: initial testing to predict mild cognitive impairment patients who convert to Alzheimer's disease. Front. Neurol. **2**, 54 (2011)

12. Parisot, S., et al.: Spectral graph convolutions for population-based disease prediction. In: Descoteaux, M., Maier-Hein, L., Franz, A., Jannin, P., Collins, D.L., Duchesne, S. (eds.) MICCAI 2017. LNCS, vol. 10435, pp. 177–185. Springer, Cham (2017). https://doi.org/10.1007/978-3-319-66179-7_21

13. Rao, N., Yu, H.-F., Ravikumar, P., Dhillon, I.S.: Collaborative filtering with graph information: consistency and scalable methods. In: Neural Information Processing Systems (NIPS), pp. 1–9 (2015)

14. Srebro, N., Rennie, J.D.M., Jaakkola, T.S.: Maximum-margin matrix factorization. In: Advances in Neural Information Processing Systems (NIPS), pp. 17:1329–17:1336 (2005)

15. Thung, K.-H., Adeli, E., Yap, P.-T., Shen, D.: Stability-weighted matrix completion of incomplete multi-modal data for disease diagnosis. In: Ourselin, S., Joskowicz, L., Sabuncu, M.R., Unal, G., Wells, W. (eds.) MICCAI 2016. LNCS, vol. 9901, pp. 88–96. Springer, Cham (2016). https://doi.org/10.1007/978-3-319-46723-8_11

16. Zhang, D., Wang, Y., Zhou, L., Yuan, H., Shen, D.: Multimodal classification of alzheimer's disease and mild cognitive impairment. NeuroImage **55**(3), 856–867 (2011)

BrainParcel: A Brain Parcellation Algorithm for Cognitive State Classification

Hazal Mogultay[(✉)] and Fatos Tunay Yarman Vural

Department of Computer Engineering, Middle East Technical University,
Ankara, Turkey
{hazal,vural}@ceng.metu.edu.tr

Abstract. In this study, we propose a novel brain parcellation algorithm, called BrainParcel. BrainParcel defines a set of supervoxels by partitioning a voxel level brain graph into a number of subgraphs, which are assumed to represent "homogeneous" brain regions with respect to a predefined criteria. Aforementioned brain graph is constructed by a set of local meshes, called mesh networks. Then, the supervoxels are obtained using a graph partitioning algorithm. The supervoxels form partitions of brain as an alternative to anatomical regions (AAL). Compared to AAL, supervoxels gather functionally and spatially close voxels. This study shows that BrainParcel can achieve higher accuracies in cognitive state classification compared to AAL. It has a better representation power compared to similar brain segmentation methods, reported the literature.

Keywords: fMRI · Brain partitioning · Mesh model

1 Introduction

Functional Magnetic Resonance Imaging (fMRI) is one of the most common imaging techniques for detecting the activation levels of human brain, during a cognitive process. fMRI measures the change of oxygen level in the brain with respect to neural activities. In principle, oxygen dependencies of neuron groups fluctuate in accordance with the activation and MRI machines can detect those changes through the scan. An intensity value is recorded at each 1–2 s for a neuron group called voxel. Each voxel is a cubic volume element around 1–2 mm^3 size. Classification of the cognitive stimulus from the voxel intensity values are called brain decoding and the pioneering studies in this area are called Multi Voxel Pattern Analysis (MVPA) [9,11]. MVPA involves recognizing the cognitive states represented by voxel intensity values of fMRI data, using machine learning techniques. A set of features is extracted from voxel intensity values recorded during each cognitive task. However, due to the large feature space formed by

© Springer Nature Switzerland AG 2018
D. Stoyanov et al. (Eds.): GRAIL 2018/Beyond MIC 2018, LNCS 11044, pp. 32–42, 2018.
https://doi.org/10.1007/978-3-030-00689-1_4

voxels (about 100,000–200,000 voxels per brain volume), dimension reduction techniques are required, such as, clustering the voxels groups into homogeneous regions.

Anatomical regions, defined by experimental neuroscience can be used as brain parcels. In most common approach, called AAL, there are 116 major regions and each region is assumed to contain voxel groups which work together. In order to reduce the dimension of the feature space, representative signals can be selected for each region or average time series can be computed within each region [1,16]. However, anatomical regions lose the subject-specific and task dependent information of brain activities. Besides, sizes of the regions vary extremely and activation levels of voxels may not be homogeneous within an anatomic region.

In order to partition the voxels into a set of homogeneous regions, well-defined clustering methods such as k-means [6,7,10], hierarchical clustering [1,4], and spectral clustering [17,20] can be used. The pros and cons of these clustering algorithms are widely studied in fMRI literature on a variety of datasets [8,18]. Some studies bring spatially close voxels together considering only the location information in analogy with the AAL [6]. Although this method improves the strict norms of AALs, it lacks the functional similarity of voxel time series, which belongs to the same regions. Recent literature reveals that functionally close voxels tend to contribute to the same cognitive task, thus, form homogeneous regions. Therefore, one needs to bring both functionally similar and spatially contiguous voxels together to define homogeneous brain regions [21]. Similarly, Wang et al. suggest to combine n-cut segmentation algorithm with simple linear iterative clustering (SLIC) [21]. Blumensath et al. use region growing for brain segmentation with functional metrics and spatial constraints between samples [3]. Bellec et al. also use region growing with functional metrics within the 26 spatial neighborhood in 3-Dimensional space [2]. Background on neuronal activity, also, supports this idea, such that physically close neurons are in chemical interaction with each other and this interaction can be interpreted as functional similarity. With these objectives in mind, many different clustering algorithms are applied to create data dependent homogeneous brain parcels. Depending on the predefined distance measure, the clustering algorithms can group spatially or functionally similar voxels under the same cluster. Craddock et al. adopt this idea and propose a brain parcellation method, in which they represent the voxels in a graph structure and used n-cut on a spatially constrained brain graph with functional edges [5]. In order to achieve spatial contiguity they connected each voxel to its 26 closest neighbors in 3D space. On the other hand, to accomplish functional homogeneity, they set edge weights of the graph to the correlation between the time series of two voxels as follows;

$$e_{i,j} = \begin{cases} corr(\mathbf{v_i}, \mathbf{v_j}) & , dist(\mathbf{v_i}, \mathbf{v_j}) \leq d_t \\ 0 & , otherwise, \end{cases} \tag{1}$$

where d_t is selected to be $\sqrt{3}$ and $corr(\mathbf{v_i}, \mathbf{v_j})$ is the Pearson Correlation between the intensity values of voxels $\mathbf{v_i}$ and $\mathbf{v_j}$. They, also, remove the edges with correlation values less than 0.5 to reduce the weak connections. Then, they define a brain graph $G = (V, E)$, where the set of voxels $V = [\mathbf{v_1}, \mathbf{v_2} \ldots \mathbf{v_N}]$ are the nodes of the graph, and $E = [e_{1,1}, e_{1,2}, \ldots e_{N,N}]$ are the edge weights computed according to Eq. 1. They partition the graph G into subgraphs by removing the edges iteratively using N-cut segmentation method using the following formula,

$$N\_cut = \frac{\sum_{\mathbf{v_i} \in A, \mathbf{v_j} \in B} e_{i,j}}{\sum_{\mathbf{v_i} \in A, \mathbf{v_n} \in V} e_{i,n}} + \frac{\sum_{\mathbf{v_i} \in A, \mathbf{v_j} \in B} e_{i,j}}{\sum_{\mathbf{v_j} \in B, \mathbf{v_n} \in V} e_{j,n}}. \tag{2}$$

As it is mentioned above, conventional MVPA methods create features sets from the selected voxel intensity values or use some averaging techniques to represent each brain region. This approach is quite restrictive to represent cognitive states. Recent studies suggest to model the relationships among voxels rather than using voxel intensity values. Ozay et al., demonstrate this idea by suggesting the Mesh Model which is a graph structure that identifies the connectivity among voxels [15].

Mesh Model (MM) represents intensity values of voxels as a weighted linear combination of its neighboring voxels, defined on a neighborhood system. The estimated weights represent the arc weights between the voxels and the voxels represents a node in the overall brain graph. A star mesh is formed around each voxel and its p neighbors, independently. In each mesh, the voxel at the center is called seed-voxel and the surrounding voxels are called neighbors. p nearest neighbors of voxel $\mathbf{v_i}$ for cognitive stimulus k are shown as $\eta^p_{\mathbf{v_i}(k)}$ and they can be selected spatially (Spatial Mesh Model - SMM) [12,14] or functionally (Functional Mesh Model - FMM) [12,13] such that, spatial neighbors has the smallest Euclidean distance with the seed-voxel whereas functional neighbors has maximum functional similarity. Meshes are formed using the full length time series for voxels, recorded during an fMRI experiment session. Assuming s measurements are taken for each cognitive stimulus, time series of a voxel $\mathbf{v_i}$ for stimulus k is an s dimensional vector shown as $\mathbf{v_i}(k) = [\mathbf{v_i}(k)^1, \mathbf{v_i}(k)^2, \ldots \mathbf{v_i}(k)^s]$. Spatial Mesh Model (SMM) selects the neighbors according to the physical distances among voxels in 3-dimensional space by Euclidean distance [12,14]. On the other hand, Functional Mesh Model (FMM), proposed by Onal et al., selects functional neighbors with the highest p-correlation values obtained by Pearson Correlation [12,13]. Afterwards, time series of the seed voxel is represented as a weighted combination of its neighbors by the following equation for each cognitive stimuli:

$$\mathbf{v_i}(k) = \sum_{\mathbf{v_j}(k) \in \eta^p_{\mathbf{v_i}(k)}} a_{i,j,k} \mathbf{v_j}(k) + \epsilon_{i,j}, \tag{3}$$

where $\eta^p_{v_i(k)}$ is the p nearest neighbors of voxel v_i for sample k and $a_{i,j,k}$ are the arc weights of the mesh network between the voxels and they are called Mesh Arc Descriptors (MADs). MADs are estimated using regularized Ridge regression method by the minimization of error term $\epsilon_{i,j}$. Concatenating each MAD for each voxel and cognitive task creates a new feature space and classification is performed on this feature space.

In this study we combine classical brain parcellation approach proposed by Craddock et al. and Mesh Model and propose a novel brain parcellation algorithm, called BrainParcel. Unlike current methods, we partition the graph formed by star meshes and partition this graph into brain regions. We show that brain partitions obtained by BrainParcel have better representation power than the partitions obtained by state of the art clustering methods and AAL in cognitive state classification problem.

Fig. 1. Overall architecture of the BrainPacel algorithm.

2   BrainParcel

Brain parcel is a brain partitioning algorithm that uses graph theoretic approaches. First, we form a brain graph by ensembling the meshes estimated around each voxel. Then, we partition this graph using n-cut segmentation algorithm. Each region is represented by the average time series of all voxels in that region. Then, these representative time series are fed to a machine learning algorithm to classify the underlying cognitive states. Figure 1 indicates the stages of suggested BrainParcel method for brain decoding problem. Each stage is explained in the following subsections.

2.1   Neighborhood System

In order to estimate a star mesh around each voxel independently, we need to define a neighborhood system. The concept of neighborhood takes an important place in this study. We inspire from the biological structure of human brain, where spatially close neurons act together by means of some electro-chemical interactions. Additionally, experimental evidence indicates that physically far apart neurons may contribute to the same cognitive process through the brain connectome. We try to utilize these observations in our brain parcellation model by defining a neighborhood system around each voxel and employ multiple connections between the neighboring voxels.

Neighborhood of the ith voxel $\mathbf{v_i}$, is defined as the set of voxels that are closest to $\mathbf{v_i}$ according to a predefined rule. Assuming p_c many neighbors around a voxel, neighborhood of $\mathbf{v_i}$ is represented by $\eta_{v_i}^{p_c}$.

Letting N be the number of voxels, we define an $N-by-N$ adjacency matrix, ND, to represent the neighborhood of voxels. Each entry of ND is calculated as follows;

$$ND(i, j) = \begin{cases} 1 & , v_j \in \eta_{v_i}^{p_c} \\ 0 & , otherwise. \end{cases} \tag{4}$$

In this study, we define two types of neighborhood, given below:

Spatial neighborhood $\eta_{v_i}^{p_c}$ is defined as the set of voxels, which has the p_c smallest Euclidean distance in 3-Dimensional space to voxel $\mathbf{v_i}$. This neighborhood system ensures resulting brain parcels to be spatially contiguous.

Functional neighborhood $\eta_{v_i}^{p_c}$ is defined as the set of voxels, which has the highest p_c-Pearson Correlation to voxel $\mathbf{v_i}$. This neighborhood system connects functionally similar voxels, even if they are physically apart from each other.

Note that, selection of the number of neighbors, p_c, and the type of the neighborhood system highly effects the rest of the steps of BrainParcel. Specifically, functional neighborhood relaxes the spatial similarity, selecting the neighboring voxels which are physically far apart. Therefore, the resulting brain parcels are not guaranteed to be spatially contiguous. It is very crucial to define a sort of balance in these two types of neighborhood, so that the resulting brain parcels consist of functionally similar and spatially contiguous voxels.

2.2 Extracting Mesh Arc Descriptors (MADs) Among Voxels

Each voxel is connected to its neighboring voxels according to one of the above neighborhood systems to form a star mesh around that voxel. The structure of star mesh depends on the type of the neighborhood system defined above. The arc weights of each local mesh are estimated by adopting the mesh model of Onal et al. [12–14]. As opposed to the current studies, we form the meshes, based on the complete time series recorded at each voxel rather than forming a different mesh for each cognitive task. This approach enables us to form a shared brain partition across all of the cognitive tasks

fMRI technique collects a time series for each voxel, when the subject is exposed to a cognitive stimulus. In the case of a block experiment design, which we have used, subjects are exposed to a stimulus for a specific time interval and the voxel time series over the entire brain volume are collected. Then, after a rest period, another stimulus is given to the subject. The time series recorded during a stimulus at i^{th} voxel is represented by the vector $\mathbf{v_i}$. Based on the idea of mesh model, we represent each $\mathbf{v_i}$ as weighted sum of other voxels in the $\eta_{v_i}^{p_c}$ neighborhood of $\mathbf{v_i}$ according to Eq. 3. Notice that Mesh Arc Descriptors (MADs) for classification are calculated per cognitive stimulus. However, we compute MADs from the entire time series of the voxels. Therefore, k index, which indicates a specific cognitive task, is removed from Eq. 3, since we compute MADs for the entire time duration of fMRI recordings. This representation is carried with a linear equation by the following formula;

$$\mathbf{v_i} = \sum_{v_j \in \eta_{v_i}^p} a_{i,j} \mathbf{v_j} + \epsilon_{i,j}. \tag{5}$$

Weights of the representation, called Mesh Arc Descriptors (MADs) are shown as $a_{i,j}$ and are estimated by Regularized Ridge Regression which minimizes the mean squared error $\epsilon_{i,j}^2$ [12–14].

2.3   Constructing a Voxel-Level Brain Graph

In order to construct a brain graph from the estimated MADs, we ensemble all the local meshes under the same graph, $G_m = (V, E_m)$. The set of nodes of this graph correspond the set of voxels $V = [\mathbf{v_1}, \mathbf{v_2}, \ldots \mathbf{v_N}]$. The set of edges corresponds to set of all MADs, $a_{i,j} \epsilon E_m$. Note that, since $a_{i,j} \neq a_{j,i}$, the graph G_m is directed. On the other hand, the graph partitioning methods, such as n-cut requires undirected graphs, in which each edge weight, $e_{i,j}$ is represented by a scalar number. In order to obtain an undirected graph from the directed graph G_m, a set of heuristic rules are used. Suppose that the mesh is formed for the voxel $\mathbf{v_i}$, and $\mathbf{v_j}$ is in the neighborhood of $\mathbf{v_i}$ with mesh arc-weight $a_{i,j}$. Edge value $e_{i,j}$ is determined, based on the following rules:

- **Case 1:** IF $\mathbf{v_i} \notin \eta_{v_j}^{p_c}$ AND $\mathbf{v_j} \in \eta_{v_i}^{p_c}$ THEN $e_{i,j} = a_{i,j}$
- **Case 2:** IF $\mathbf{v_i} \in \eta_{v_j}^{p_c}$ AND $\mathbf{v_j} \in \eta_{v_i}^{p_c}$ THEN this case requires further analysis. Assuming highly correlated voxels should have a stronger edge between them, we employ the following thresholding method;
  IF $corr(v_i, v_j) \geq 0$, THEN $e_{i,j} = max(a_{i,j}, a_{j,i})$
  IF $corr(v_i, v_j) < 0$, THEN $e_{i,j} = min(a_{i,j}, a_{j,i})$.

Above rules prune the directed graph G_m to an undirected graph G to be partitioned for obtaining homogeneous brain regions, called supervoxels.

2.4   Graph Partitioning for Obtaining Supervoxels

After constituting the brain graph G, n-cut segmentation method is used for clustering this graph. N-cut is a graph partitioning algorithm which carries a graph cut method on a given undirected graph. Given G, n-cut cuts the edges one by one in an iterative manner. With each cut, the graph is split into two smaller connected components. Letting N be the number of voxels, n-cut method requires the representative graph G, which is actually an $N - by - N$ adjacency matrix explained in the previous sections. The number of intended brain parcels is set to $\mathbb{C}$. With graph cut operations, graph is split into C connected components where $C \leq \mathbb{C}$. Each sample is a member of one of this clusters and assigned with a cluster index. In other words, n-cut method returns an $1 - by - N$ dimensional vector $\mathbb{L}_C = [l_1^c, l_2^c, \ldots, l_N^c]$ where each l_i^c is a number between 1 and C. The n-cut method, as applied to undirected graph G is called BrainParcel. The output of this algorithm yields a set of supervoxels, which are homogeneous with respect to the subgraphs of mesh network.

Recall that, anatomical regions (AAL) produce an experimentally neuroscientific parcellation of the brain. In order to compare the brain decoding performances, we conducted our experiment, where we form mesh network for both among anatomical regions and the network formed among supervoxels obtained at the output of BrainParcel. There are 116 basic brain regions in AAL and each voxel resides in one and only one region. Let us represent the anatomically defined region indices of voxels with $\mathbb{L}_\mathbb{A} = [l_1^a, l_2^a, \ldots, l_N^a]$, in order to avoid confusions. Notice that, with $\mathbb{L}_{\mathbb{A}}$ we skip all of the brain parcellation steps. Also, let us call $\mathbb{L} = [l_1, l_2, \ldots, l_N]$ to all kinds of brain segmentations; in our case it means $\mathbb{L} \supset (\mathbb{L}_{\mathbb{C}} \cap \mathbb{L}_{\mathbb{A}})$.

2.5   Representation of Supervoxels

We need to calculate a representative signal for each supervoxel. For this purpose, we take an average among the time series of voxels within each supervoxel. With C supervoxels, we calculate set of vectors $U = [\mathbf{u_1}, \mathbf{u_2}, \ldots, \mathbf{u_C}]$, where each u_i is the representative vector of supervoxel i and they are calculated as follows;

$$\mathbf{u_i} = \frac{\sum_{l_j==i} \mathbf{v_j}}{\sum_{l_j==i} 1}. \tag{6}$$

In the dataset on which we have performed our experiments, six measurements were taken for each cognitive stimulus. Assuming K stimuli were shown to each subject, time series of each voxel has a length $\mathbb{K} = 6K$. Therefore, at the output of the clustering algorithm we construct a data matrix U of size $C - by - \mathbb{K}$, where each row represents a feature, and each column corresponds to a cognitive stimulus.

2.6   Constructing Supervoxel-Level Brain Graph

The original area of utilization of the mesh model was to model the relationships among voxels and use this relationship for decoding the cognitive processes. Both spatial and functional neighborhoods were considered, and their representation powers were demonstrated by relatively high recognition performances compared to the available state of the art network models. Specifically, Functional Mesh Model (FMM) outperform most of the MVPA and Spatial Mesh Model (SMM) results. Therefore, we use FMM for classifying the cognitive states.

Data matrix U, defined in the previous section, is feed into the FMM algorithm. Each supervoxel $\mathbf{u_i}$ is represented by linear combination of its functional neighbors, the arc weights are estimated at each mesh using Ridge Regression for each cognitive stimulus. Recall that fMRI collects multiple measurements during the time course of each cognitive stimulus. In our dataset 6 measurements are collected for each stimulus, and $\mathbf{u_i}$ is a vector of length $\mathbb{K} = 6K$ for K stimuli. Let us represent the vector of the stimulus k by $\mathbf{u_i}(\mathbf{k})$. First, functionally closest p_m neighbors of $\mathbf{u_i}(\mathbf{k})$; $\eta_{u_i(k)}^{p_m}$, are selected from the supervoxels $\mathbf{u_j}$ which has the highest correlation with supervoxel $\mathbf{u_i}$ according to Pearson Correlation. Then, the mean square error $E(\epsilon_{i,j}^2)$ is minimized to estimate $a_{i,j,k}$ of the Eq. 3. Estimated MADs are concatenated so that they represent a more powerful feature space compared to the raw fMRI signal intensity values that is used in MVPA studies. We concatenate all the MADs and represent the stimulus in a feature space formed by MADs.

2.7   Classification

MADs estimated at supervoxel-level, are concatenated under a feature vector for classifying the cognitive states. 6 fold cross validation schema is applied on the dataset, where at each fold, one run is reserved from the data as a test set. Logistic regression is used for classification.

3   Experiments

3.1   Dataset

In this study, we use a dataset called "Object Experiment". This dataset is recorded by the team of ImageLab of METU members at Bilkent University UMRAM Center. It consists of 4 subjects in the age of twenties. Each subject is shown various bird and flower pictures. In between those stimuli, simple mathematics questions are shown as transition. There are 6 runs in the experiment and in each run, 36 pictures are shown to each subject. Thus, there are total of 216 samples. Number of samples are balanced for the two classes (bird and flower). Preprocessing of the dataset is carried with the SPM toolbox and the number of voxels is decreased to 20,000 for each subject. Also, there are 116 labeled anatomical regions, defined under MNI coordinate system [19]. We provide experimental results, where each given accuracy is the output of a six fold cross validation. Recall that, each subject is given 6 runs of stimuli. At each fold, we split a run for testing and use the other 5 runs for training. The reported accuracies are the average of these 6 folds for each subject.

3.2   Comparative Results

In this section, we provide a comparison between BrainParcel and the parcellation algorithm suggested by Craddock et al. Table 1 shows the classification performances for various number of parcels. The results are reported after optimizing the mesh sizes empirically. Recall that, functionally constrained systems that construct the graph with *Functional_ND* neighborhood system does not provide any spatial integrity within the brain parcels, since the brain graph is not formed on these grounds. On the other hand, spatially constrained systems provide both spatial continuity and functional homogeneity since the brain graph is formed by spatial restrictions and edges are weighted in terms of functional connectivity.

In Table 1, the first and third column give the best results for the brain parcellation method suggested by Craddock et al. (called classical, in the Table), and the other two gives the results obtained by BrainParcel that we have proposed. Each row of this table gives the results for a different number of supervoxels (SV). Notice that spatially constrained BrainParcel gives the best classification performances in the overall schema.

These results point to the idea that, in order to achieve better representational power for cognitive state classification, one needs spatially contiguous and functionally homogeneous brain parcels, which is accomplished by spatial BrainParcel. Moreover, recall that we have offered BrainParcel as an alternative to AAL, which has 116 basic anatomic regions and gives 53% performance on average. A compatible parcellation scheme consists of 100 super voxels, where, Spatial BrainParcel results in higher classification accuracies compared to the other methods.

Table 1. Overall 2-class classification accuracy acquired from the MADs constructed among super voxels and method suggested by Craddock et al. (called, classical). These results suggest that Spatial BrainParcel gives higher performances, since it provides spatial continuity and functional homogeneity within each brain parcel.

# of SV	Spatial constraints		Functional constraints	
	Classical	BrainParcel	Classical	BrainParcel
100	67.79	**74.00**	70.63	73.29
250	72.96	**78.71**	76.79	77.54
500	75.46	**79.42**	77.33	77.54
750	77.46	78.04	**79.38**	77.54
1000	78.08	78.83	79.96	**80.08**

4   Conclusion

In this study, we offer a brain parcellation methodology, which combines the spatial and functional connectivity of brain on a novel graph representation. This approach offers a better alternative to the current brain parcellation methods in the literature [8,18], for brain decoding problems. BrainParcel uses spectral clustering methods, which represents the voxel space as a graph formed by mesh model. Common studies compute the edge weights of the brain graph as the pairwise correlation between voxels, whereas we computed the edge weights by estimating them using the mesh model among a group of voxels. Then, brain graph is partitioned with n-cut segmentation method to generate supervoxels.

As suggested, using task dependent brain parcellation methods enable better brain decoding performances compared to anatomical regions. Moreover, it is demonstrated that functional connectivity, united with the spatial contiguity is the best approach to represent homogeneous brain regions.

Also, results show that using the MADs of the mesh model for classification, improves the brain decoding performances in all of the experiment setups.

Our study reveals that mesh model not only improves the classification performance, but also creates a brain graph, where the nodes represent homogeneous super voxels with a better representation power for brain decoding. Although the performance increase looks relatively small, when the large size of the data set is considered, the performance boost becomes quite meaningful.

In the future, experimental set up can be refined for parameter selection.

Acknowledgement. This project is supported by TUBITAK under grant number 116E091. We thank UMRAM Center of Bilkent University for opening their facilities to collect fMRI dataset. We also thank to Dr. Itir Onal Ertugrul and Dr. Orhan Firat for their contribution and effort of data collection.

References

1. Alkan, S., Yarman-Vural, F.T.: Ensembling brain regions for brain decoding. In: 37th Annual International Conference of the IEEE Engineering in Medicine and Biology Society (EMBC), pp. 2948–2951. IEEE (2015)
2. Bellec, P., et al.: Identification of large-scale networks in the brain using fMRI. Neuroimage **29**(4), 1231–1243 (2006)
3. Blumensath, T., et al.: Spatially constrained hierarchical parcellation of the brain with resting-state fMRI. Neuroimage **76**, 313–324 (2013)
4. Cordes, D., Haughton, V., Carew, J.D., Arfanakis, K., Maravilla, K.: Hierarchical clustering to measure connectivity in fMRI resting-state data. Magnetic Reson. Imaging **20**(4), 305–317 (2002)
5. Craddock, R.C., James, G.A., Holtzheimer, P.E., Hu, X.P., Mayberg, H.S.: A whole brain fMRI atlas generated via spatially constrained spectral clustering. Hum. Brain Mapp. **33**(8), 1914–1928 (2012)
6. Flandin, G., Kherif, F., Pennec, X., Malandain, G., Ayache, N., Poline, J.-B.: Improved detection sensitivity in functional MRI data using a brain parcelling technique. In: Dohi, T., Kikinis, R. (eds.) MICCAI 2002. LNCS, vol. 2488, pp. 467–474. Springer, Heidelberg (2002). https://doi.org/10.1007/3-540-45786-0_58
7. Flandin, G., Kherif, F., Pennec, X., Riviere, D., Ayache, N., Poline, J.B.: Parcellation of brain images with anatomical and functional constraints for fmri data analysis, pp. 907–910 (2002)
8. Liao, T.W.: Clustering of time series data'a survey. Pattern Recogn. **38**(11), 1857–1874 (2005)
9. Mitchell, T., et al.: Predicting human brain activity associated with the meanings of nouns. Science **320**(5880), 1191–1195 (2008)
10. Moğultay, H., Alkan, S., Yarman-Vural, F.T.: Classification of fMRI data by using clustering. In: 23th Signal Processing and Communications Applications Conference, SIU, pp. 2381–2383. IEEE (2015)
11. Norman, K.A., Polyn, S.M., Detre, G.J., Haxby, J.V.: Beyond mind-reading: multivoxel pattern analysis of fMRI data. Trends Cogn. Sci. **10**(9), 424–430 (2006)
12. Onal, I., Ozay, M., Mizrak, E., Oztekin, I., Yarman-Vural, F.T.: A new representation of fMRI signal by a set of local meshes for brain decoding. IEEE Trans. Sig. Inf. Process. Over Netw. (2017). https://doi.org/10.1109/TSIPN.2017.2679491
13. Onal, I., Ozay, M., Yarman-Vural, F.T.: Functional mesh model with temporal measurements for brain decoding. In: 37th Annual International Conference of the IEEE Engineering in Medicine and Biology Society (EMBC), pp. 2624–2628. IEEE (2015)
14. Onal, I., Ozay, M., Yarman-Vural, F.T.: Modeling voxel connectivity for brain decoding. In: International Workshop on Pattern Recognition in NeuroImaging (PRNI), pp. 5–8. IEEE (2015)
15. Ozay, M., Öztekin, I., Öztekin, U., Yarman-Vural, F.T.: Mesh learning for classifying cognitive processes (2012). arXiv preprint arXiv:1205.2382
16. Richiardi, J., Eryilmaz, H., Schwartz, S., Vuilleumier, P., Van De Ville, D.: Decoding brain states from fMRI connectivity graphs. Neuroimage **56**(2), 616–626 (2011)

17. Shen, X., Papademetris, X., Constable, R.T.: Graph-theory based parcellation of functional subunits in the brain from resting-state fMRI data. Neuroimage **50**(3), 1027–1035 (2010)
18. Thirion, B., Varoquaux, G., Dohmatob, E., Poline, J.B.: Which fMRI clustering gives good brain parcellations? Front. Neurosci. **8** (2014)
19. Tzourio-Mazoyer, N., et al.: Automated anatomical labeling of activations in SPM using a macroscopic anatomical parcellation of the MNI MRI single-subject brain. Neuroimage **15**(1), 273–289 (2002)
20. Van Den Heuvel, M., Mandl, R., Pol, H.H.: Normalized cut group clustering of resting-state fMRI data. PloS One **3**(4), e2001 (2008)
21. Wang, J., Wang, H.: A supervoxel-based method for groupwise whole brain parcellation with resting-state fMRI data. Front. Hum. Neurosci. **10** (2016)

Modeling Brain Networks with Artificial Neural Networks

Baran Baris Kivilcim[1(✉)], Itir Onal Ertugrul[2], and Fatos T. Yarman Vural[1]

[1] Department of Computer Engineering, Middle East Technical University,
Ankara, Turkey
{baran.kivilcim,vural}@ceng.metu.edu.tr
[2] Robotics Institute, Carnegie Mellon University, Pittsburgh, PA, USA
iertugru@andrew.cmu.edu

Abstract. In this study, we propose a neural network approach to capture the functional connectivities among anatomic brain regions. The suggested approach estimates a set of brain networks, each of which represents the connectivity patterns of a cognitive process. We employ two different architectures of neural networks to extract directed and undirected brain networks from functional Magnetic Resonance Imaging (fMRI) data. Then, we use the edge weights of the estimated brain networks to train a classifier, namely, Support Vector Machines (SVM) to label the underlying cognitive process. We compare our brain network models with popular models, which generate similar functional brain networks. We observe that both undirected and directed brain networks surpass the performances of the network models used in the fMRI literature. We also observe that directed brain networks offer more discriminative features compared to the undirected ones for recognizing the cognitive processes. The representation power of the suggested brain networks are tested in a task-fMRI dataset of Human Connectome Project and a Complex Problem Solving dataset.

Keywords: Brain graph · Brain decoding · Neural networks

1 Introduction

Brain imaging techniques, such as, functional Magnetic Resonance Imaging (fMRI) have facilitated the researches to understand the functions of human brain using machine learning algorithms [14,15,20,25]. In traditional approaches, such as Multi-Voxel Pattern Analysis (MVPA), the aim was to discriminate cognitive tasks from the fMRI data itself without forming brain graphs and considering relationship between nodes of graphs. Moreover, Independent Component Analysis (ICA) and Principal Component Analysis (PCA) have been applied to obtain better representations. In addition to feature extraction methods, General Linear Model (GLM) and Analysis of Variance (ANOVA) have been used to select important voxels [20]. None of these approaches take into account the

© Springer Nature Switzerland AG 2018
D. Stoyanov et al. (Eds.): GRAIL 2018/Beyond MIC 2018, LNCS 11044, pp. 43–53, 2018.
https://doi.org/10.1007/978-3-030-00689-1_5

massively connected network structure of the brain [3,4,12,22,26]. Recently, use of deep learning algorithms have also emerged in several studies [7–9] to classify cognitive states. Most of these studies mainly focus on using deep learning methods to extract better representations from fMRI data for brain decoding.

Several studies form brain graphs using voxels or anatomical regions as nodes and estimate the edge weights of brain graphs with different approaches. Among them, Richiardi et al. [21] have created undirected functional connectivity graphs in different frequency subbands. They have employed Pearson correlation coefficient between responses obtained from all region pairs as edge weights and use these edge weights to perform classification in an audio-visual experiment. Brain graphs, constructed using pairwise correlations and mutual information as edge weights, have been used to investigate the differences in networks of healthy controls and patients with Schizophrenia [11] or Alzheimer's disease [10,13]. Yet, these studies consider only pairwise relationships while estimating the edge weights and ignore the locality property of the brain.

Contrary to pairwise relationships, a number of studies have estimated the relationships among nodes within a local neighborhood. Ozay et. al. [19] and Firat et al. [6] have formed local meshes around nodes and constructed directed graphs as ensembles of local meshes. They have applied Levinson-Durbin recursion [24] to estimate the edge weights representing the linear relationship among voxels and have used these weights to classify the category of words in a working memory experiment. Similarly, Alchihabi et al. [2] have applied Levinson-Durbin recursion to estimate the edge weights of local meshes of dynamic brain network for every brain volume in Complex Problem Solving task and have explored activation differences between sub-phases of problem solving. While these studies conserve the locality in the brain, construction of a graph for every time instant discards temporal relationship among nodes of the graph. Onal et al. [17,18] have formed directed brain graphs as ensemble of local meshes. They have estimated the relationships among nodes within a time period considering the temporal information using ridge regression. Since the spatially neighboring voxels are usually correlated, linear independence assumption of features required for closed form solution to the estimation of linear relationship among voxels is violated. This may result in large errors and inadequate representation. Since the aforementioned studies form local meshes around each node separately, associativity is ignored in the resulting brain graphs.

In this study, we propose two brain network models, namely, directed and undirected Artificial Brain Networks to model the relationships among anatomical regions within a time interval using fMRI signals. In both network models, we train an artificial neural network to estimate the time series recorded at node which represent an anatomic region by using the rest of the time series recorded in the remaining nodes. In our first neural network architecture, called directed Artificial Brain Networks (dABN), global relationships among nodes are estimated without any constraint whereas in our second architecture of undirected Artificial Brain Networks (uABN), we apply a weight sharing mechanism to ensure undirected functional connections.

We test the validity of our dABN and uABN in two fMRI datasets and compare the classification performances to the other network models available in the literature. First, we employ the Human Connectome Project (HCP), task-fMRI (tfMRI) dataset, in which the participants were required to complete 7 different mental tasks. The second fMRI dataset contains fMRI scans of subjects solving Tower of London puzzle and has been used to study regional activations of Complex Problem Solving [2,16]. The task recognition performances of the suggested Artificial Brain Networks are significantly greater than the ones obtained with state of the art functional connectivity methods.

2   Extraction of Artificial Brain Networks

In this section, we explain how we estimate the edge weights of directed and undirected brain networks using artificial neural networks. Throughout this study, we represent a brain network by $G = (V, E)$, where $V = \{v_1, v_2, v_3, \ldots, v_M\}$, denotes the vertices of the network, which represent $M = 90$ anatomical brain regions, $R = \{r_1, r_2, r_3, \ldots, r_M\}$. The attribute of each node is the average time series of BOLD activations. The average BOLD activation of an anatomical region r_i at time t is denoted with $b_{i,t}$. We use all anatomical regions defined by Anatomical Atlas Labeling (AAL) [23], except for the ones residing in Cerebellum and Vermis. We represent the edges of the brain network by $E = \{e_{i,j} | \forall v_i, v_j \in V, i \neq j\}$. The weights of edges depend on the estimation method. We denote the adjacency matrix which consists of the edge weights, as A, where $a_{i,j}$ represents the weight of edge from v_i to v_j, when the network is directed. When the network is undirected the weight of the edge formed between v_i and v_j is $a_{i,j} = a_{j,i}$. Sample representations of directed and undirected brain networks are shown in Figs. 1 and 2, respectively.

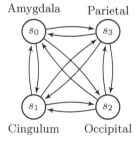

Fig. 1. A directed brain network.

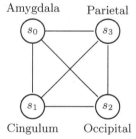

Fig. 2. An undirected brain network.

We temporally partition the fMRI signal into chunks with length L recorded during each cognitive process. The fMRI time series at each chunk is used to estimate a network to represent the spatio-temporal relationship among anatomic regions. Then, the cognitive process k of subject s is described as a consecutive list (T_k^s) of brain networks, formed for each chunk within time interval $[t, t+L]$, where $T_k^s = \{G_1, G_2, \ldots, G_{C_k}\}$. Note that, C_k is the number of chunks obtained for cognitive process k and equals to $\lfloor N_k/L \rfloor$, where N_k denotes the number of measurements recorded for cognitive process k. Since we obtain a different network for each duration of length L for a cognitive process of length N_k, this approach estimates a dynamic network for the cognitive process, assuming that N_k is sufficiently large.

For a given time interval $[t, t+L]$, weights of incoming edges to vertex v_i is defined by an M dimensional vector, $\bar{\mathbf{a}}_\mathbf{i} = [a_{i,1}, a_{i,2} \ldots a_{i,M}]$. Note that the ith entry $a_{i,i} = 0$, which implies that a node does not have an edge value into itself. These edge weights define the linear dependency of activation, $b_{i,t}$, of region r_i at time t to the activations of the remaining regions, $b_{j,t}$ for a time interval $t' \in \{t, t+L\}$

$$b_{i,t'} = \sum_{j \neq i, j=1}^{M} a_{i,j} b_{j,t'} + \epsilon_{t'} = \hat{b}_{i,t'} + \epsilon_{t'} \qquad \forall t' \in \{t, t+L\} \qquad (1)$$

where $\hat{b}_{i,t'}$ is the estimated value of $b_{i,t'}$ at time t' with error rate $\epsilon_{t'}$, which is the difference between the real and estimated activation. Note that each node is connected to the rest of $M-1$ nodes each of which corresponding to anatomic regions.

2.1   Directed Artificial Brain Networks (dABN)

In fully connected directed networks, we define two distinct edges between all pairs of vertices, $E = \{e_{i,j}, e_{j,i} | v_i, v_j \in V, i \neq j\}$ where $e_{i,j}$ denotes an edge from v_i to v_j. The weights of the edge pairs are not to be symmetrical, $a_{i,j} \neq a_{j,i}$.

The neural network we design to estimate edge weights consists of an input layer and an output layer. For every edge in the brain network, we have an equivalent weight in the neural network, such that weight between $input_i$ and $output_j$, $w_{i,j}$ is assumed to be an estimate for the weight, $a_{i,j}$ of the edge from v_i to v_j, in the artificial brain network.

We employ a regularization term λ to increase generalization capability of the model and minimize the expected value of sum of squared error through time. Loss of an output node $output_i$ is defined as,

$$Loss(output_i) = E((b_{i,t'} - \sum_{j \neq i, j=1}^{M} w_{i,j} b_{j,t'})^2) + \lambda \mathbf{w}_i^T \mathbf{w}_i, \qquad (2)$$

where $w_{i,j}$ denotes the neural network weight between $input_i$ and $output_j$ and $E(.)$ is the expectation operator taken over time interval $[t, t+L]$. For each training step of the neural network, e, gradient descent is applied for the optimization of the weights as in Eq. (3) with empirically chosen learning rate, α. The whole system is trained for an empirically selected number of epochs (Fig. 3).

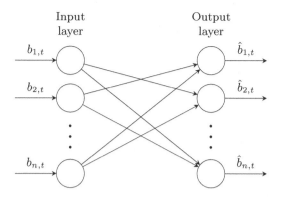

Fig. 3. Directed Artificial Brain Network architecture.

$$w_{i,j}^{(e)} \leftarrow w_{ij}^{(e-1)} - \alpha \frac{\partial Loss(output_i)}{\partial w_{i,j}}. \tag{3}$$

After training, the weights of neural network are assigned to edge weights of the corresponding brain graph, $a_{i,j} \leftarrow w_{i,j}, \forall_{i,j}$.

2.2   Undirected Artificial Brain Network (uABN)

In undirected brain networks, similar to directed brain network, we define double connections between every pair of vertices $E = \{e_{i,j}, e_{j,i}|v_i, v_j \in V, i \neq j\}$. However, in order to make the network undirected, we must satisfy the constraint that twin (opposite) edges have the equal weights, $a_{i,j} = a_{j,i}$. In order to assure th's property in the neural network explained in the previous section, we use a weight sharing mechanism and keep the weights of the twin (opposite) edges in the neural network equal through the learning process, such that $w_{i,j} = w_{j,i}$. The proposed architecture is shown at Fig. 4.

We use Eq. (2) for undirected Artificial Brain Networks. The weight matrix of uABN is initialized symmetrically, $w_{i,j} = w_{j,i}$ and in order to satisfy the symmetry constraint through training epochs, we define the following update rule for the weights, $w_{i,j}$ and $w_{j,i}$ at epoch e.

$$w_{i,j}^{(e)} = w_{j,i}^{(e)} \leftarrow w_{i,j}^{(e-1)} - \frac{1}{2}\alpha \left[\frac{\partial Loss(output_i)}{\partial w_{i,j}} + \frac{\partial Loss(output_j)}{\partial w_{i,j}} \right]. \tag{4}$$

Again, after an empirically determined number of epochs, the weights of edges in the undirected graph is assigned to the neural network weights, $a_{i,j} \leftarrow w_{i,j}$.

2.3   Baseline Methods

In this subsection, we briefly describe the popular methods that have been used to build functional connectivity graphs, in order to provide some comparison for the suggested Artificial Brain Network.

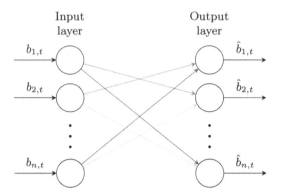

Fig. 4. Neural network structure to create undirected Artificial Brain Networks (connections with the same colors are shared).

Pearson Correlation. In their work, Richiardi et al. [21] defined the functional connectivity between two anatomic regions as pair-wise Pearson correlation coefficients computed between the average activations of these regions in a time interval. The edge weights are calculated by,

$$\rho_{b_{i,t,L}, b_{j,t,L}} = \frac{cov(\mathbf{b_{i,t,L}}, \mathbf{b_{j,t,L}})}{\sigma_{\mathbf{b_{i,t,L}}} \sigma_{\mathbf{b_{j,t,L}}}}, \tag{5}$$

where $\mathbf{b_{i,t,L}} = [b_{i,t}, b_{i,(t+1)}, \dots, b_{i,(t+L)}]$ represents the average time series of BOLD activations of region i between time t and $t + L$, $cov()$ defines the covariance, and σ_s represents the standard deviation of time series s. This approach assumes that the pair of similar time series represent the same cognitive process measured by fMRI signals.

Closed Form Ridge Regression. In order to generate brain networks with the method proposed in [18], we estimate the activation of a region from the activations of its neighboring regions in a time interval $[t, t + L]$. We minimize the loss function in Eq. 2 using closed form solution for ridge regression. The loss function is minimized with respect to the edge weights outgoing from a vertex v_i, $\bar{a}_i = [a_{i,1}, a_{i,2} \dots a_{i,M}]$ and the following closed form solution of ridge regression is obtained:

$$\mathbf{\bar{a}_i} = (\mathbf{B}^T\mathbf{B} + \lambda\mathbf{I})^{-1}\mathbf{B}^T\mathbf{b_{i,t,L}}, \tag{6}$$

where $\mathbf{B}$ is an $L \times (M - 1)$ matrix, whose columns consist of the average BOLD activations of anatomic regions except for the region r_i in the time interval $[t, t + L]$ such that column j of matrix $\mathbf{B}$ is $\mathbf{b_{j,t,L}}$. $\lambda \in R$ represents the regularization parameter.

3   Experiments and Results

In order to examine the representation power of the suggested Artificial Brain Networks, we compare them with the baseline methods, presented in the previous subsection, on two different fMRI dataset. The comparison is done by measuring the cognitive task classification performances of all the models.

3.1   Human Connectome Project (HCP) Experiment

In Human Connectome Project dataset, 808 subjects attended 7 sessions of fMRI scanning in each of which the subjects were required to complete a different cognitive task with various durations, namely, Emotion Processing, Gambling, Language, Motor, Relational Processing, Social Cognition, and Working Memory. We aim to discriminate these 7 tasks using the edge weights of the formed brain graphs.

In the experiments, the learning rate α was empirically chosen as $\alpha = 10^{-5}$ for both dABN and uABN and window size is chosen as $L = 40$. We tested the directed and undirected Artificial Brain Networks and Ridge Regression method using various λ values. Since computation of Pearson correlation does not require any hyper parameter estimation, a single result is obtained for the Pearson correlation method.

After estimating the Artificial Brain Networks and forming the feature vectors from edge weights of the brain networks, we performed within-subject and across-subject experiments using Support Vector Machines with linear kernel. During the within-subjects experiments, we performed 3-fold cross validation using only the samples of a single subject. Table 1 shows the average of within-subject experiment results over 807 subjects, when the classification is performed using a single subject brain network of 7 tasks. During the across-subject experiments, we performed 3-fold cross validation using the samples obtained from 807 subjects. For each fold we employed the samples from 538 subjects to train and 269 subject to test the classifier. Table 2 shows the across-subject experiment results.

Table 1. Within-subject performances of brain networks on HCP dataset.

λ	Pearson corr.		Ridge reg.		dABN		uABN	
	Mean	Std	Mean	Std	Mean	Std	Mean	Std
0	0.7194	0.16	-	-	**0.7435**	0.13	0.5918	0.13
32	0.7194	0.16	0.7957	0.11	**0.9133**	0.08	**0.913**	0.08
64	0.7194	0.16	0.8304	0.11	**0.9406**	0.07	**0.9402**	0.07
128	0.7194	0.16	0.8377	0.11	**0.9463**	0.06	**0.9462**	0.07
256	0.7194	0.16	0.8119	0.12	**0.9313**	0.08	**0.9307**	0.08
512	0.7194	0.16	0.7462	0.13	**0.8852**	0.1	**0.8849**	0.1

Table 2. Across-subject performances of brain networks on HCP dataset.

λ	Pearson corr.		Ridge reg.		dABN		uABN	
	Mean	Std	Mean	Std	Mean	Std	Mean	Std
0	**0.7524**	0.01	-	-	0.6654	0.01	0.5681	0.01
32	0.7524	0.01	0.8027	0.01	**0.8153**	0.00	0.8123	0.00
64	0.7524	0.01	0.8223	0.00	**0.8312**	0.01	0.8297	0.01
128	0.7524	0.01	0.8370	0.01	**0.8401**	0.01	0.8393	0.01
256	0.7524	0.01	**0.8461**	0.01	0.8410	0.01	0.8406	0.00
512	0.7524	0.01	**0.8466**	0.01	0.8357	0.01	0.8357	0.01

Table 1 shows that in within subject experiments our methods, dABN and uABN, have the best performances in classifying the cognitive task under different λ values, furthermore performances of directed networks are slightly better than undirected ones. It can be observed that as λ increases, generalization of our models also increase up to $\lambda = 128$.

Table 2 shows that our methods outperforms the others within a range of lambdas, $\lambda = \{32, 64, 128\}$. Pearson Correlation results in the best accuracy when no regularization is applied to Artificial Brain Networks. Closed Form Ridge Regression solution offers more discriminative power in higher λ values.

3.2 Tower of London (TOL) Experiment

We also test the validity of the suggested Artificial Brain Network on a relatively more difficult fMRI dataset, recorded when the subjects solve Tower of London (TOL) problem. TOL is a puzzle game which has been used to study complex problem solving tasks in human brain. TOL dataset used in our experiments contains fMRI measurements of 18 subjects attending 4 session of problem solving experiment. In the fMRI experiments, subjects were asked to solve 18 different puzzles on computerized version of TOL problem [16]. There are two labeled subtask of problem solving with varying time periods namely, planning and execution phases.

As the nature of the data is not compatible with a sliding window approach and the dimensionality is too high for a computational model, in the study of Alchihabi et al. [1], a series of preprocessing steps were suggested for the TOL dataset. In this study, we employ the first two steps of their pipeline. In the first step called *voxel selection and regrouping*, a feature selection method is applied on time series of voxels to select the "important" ones. Then, the activations of the selected voxels in the same region are averaged to obtain the activity of corresponding region. As a result, a more informative and lower dimensional representation is achieved. In the second step, bi-cubic spline interpolation is applied to every consecutive brain volumes and a number of new brain volumes are inserted between two brain volumes to increase temporal resolution. For the details of interpolation, refer to [1]. In this study, the optimal number of volumes inserted between two consecutive brain volumes are found empirically and it is set to 4. Therefore, the time resolution of the data is increased four times.

We applied the above-mentioned preprocessing steps to all of the 72 sessions in the dataset. After the voxel selection phase, number of regions containing selected voxels is much less than 116 regions. Note that, we discard regions located in Cerebellum and Vermis. Window size for this dataset was set to $L = 5$, since there are at least 5 measurements for every sub-phase after the interpolation. The neural network parameters used in our experiments are $\alpha = 10^{-6}$ and $\#epochs = 10$. Table 3 shows the mean and standard deviation of classification accuracies obtained with our method and the base-line methods. Similar to HCP experiments, we slided non-overlapping windows on the measurements and we performed 3-fold cross validation during TOL experiments.

Table 3. Across-subject performances of mesh networks on TOL dataset.

λ	Pearson corr.		Ridge reg.		dABN		uABN	
	Mean	Std	Mean	Std	Mean	Std	Mean	Std
0	0.6119	0.09	-	-	**0.8914**	0.11	0.8499	0.12
32	0.6119	0.09	0.6688	0.10	**0.8913**	0.11	0.8499	0.12
64	0.6119	0.09	0.6651	0.10	**0.8914**	0.11	0.8499	0.12
128	0.6119	0.09	0.6679	0.10	**0.8906**	0.11	0.8499	0.12
256	0.6119	0.09	0.6685	0.10	**0.8905**	0.11	0.8500	0.12
512	0.6119	0.09	0.6705	0.10	**0.8912**	0.11	0.8498	0.12

Table 3 shows that using Artificial Brain Networks gives better performances than using Pearson Correlation and Closed Form Ridge Regression methods in classifying sub-phases of complex problem solving under various regularization parameters. We observe that decoding performances of directed brain networks outperforms those of undirected brain networks.

4   Discussion and Future Work

In this study, we introduce a network representation of fMRI signals, recorded when the subjects perform a cognitive task. We show that the suggested Artificial Brain Network estimated from the average activations of anatomic regions using an artificial neural network leads to a powerful representation to discriminate cognitive processes. Compared to the brain networks obtained by ridge regression, the suggested Artificial Brain Network achieves more discriminative features. The success of the suggested brain network can be attributed to the iterative nature of the neural network algorithms to optimize the loss function, which avoids the singularity problems of Ridge Regression.

In most of the studies, it is customary to represent functional brain connectivities as an undirected graphs. However, in this study, we observe that the directed network representations capture more discriminative features compared to the undirected ones in brain decoding problems.

In this study, we consider complete brain graphs where all regions are assumed to have connections to each other. A sparser brain representation can be computationally more efficient and neuro-scientifically more accurate. As a future work, we aim to estimate more efficient brain network representations by employing some sparsity parameters in the artificial neural networks.

It is well-known that brain processes the information in various frequency bands. [5,21] applied discrete wavelet transform before creating connectivity graphs. A similar approach can be taken for a more complete temporal information in brain decoding problems.

Acknowledgment. The work is supported by TUBITAK (Scientific and Technological Research Council of Turkey) under the grant No: 116E091. We also thank Sharlene Newman, from Indiana University, for providing us the TOL dataset.

References

1. Alchihabi, A., Kivilicim, B.B., Ekmekci, O., Newman, S.D., Vural, F.T.Y.: Decoding cognitive subtasks of complex problem solving using fMRI signals. In: 2018 26th Signal Processing and Communications Applications Conference (SIU). IEEE (2018)
2. Alchihabi, A., Kivilicim, B.B., Newman, S.D., Vural, F.T.Y.: A dynamic network representation of fMRI for modeling and analyzing the problem solving task. In: 2018 IEEE 15th International Symposium on Biomedical Imaging (ISBI 2018), pp. 114–117. IEEE (2018)
3. Calhoun, V.D., Adali, T., Hansen, L.K., Larsen, J., Pekar, J.J.: ICA of functional MRI data: an overview. In: Proceedings of the International Workshop on Independent Component Analysis and Blind Signal Separation. Citeseer (2003)
4. Calhoun, V.D., Liu, J., Adalı, T.: A review of group ICA for fMRI data and ICA for joint inference of imaging, genetic, and ERP data. Neuroimage **45**(1), S163–S172 (2009)
5. Ertugrul, I.O., Ozay, M., Vural, F.T.Y.: Hierarchical multi-resolution mesh networks for brain decoding. Brain Imaging Behav. 1–17 (2016)
6. Fırat, O., Özay, M., Önal, I., Öztekiny, İ., Vural, F.T.Y.: Functional mesh learning for pattern analysis of cognitive processes. In: 2013 12th IEEE International Conference on Cognitive Informatics & Cognitive Computing (ICCI* CC), pp. 161–167. IEEE (2013)
7. Firat, O., Oztekin, L., Vural, F.T.Y.: Deep learning for brain decoding. In: 2014 IEEE International Conference on Image Processing (ICIP), pp. 2784–2788. IEEE (2014)
8. Kawahara, J., et al.: BrainNetCNN: convolutional neural networks for brain networks; towards predicting neurodevelopment. NeuroImage **146**, 1038–1049 (2017)
9. Koyamada, S., Shikauchi, Y., Nakae, K., Koyama, M., Ishii, S.: Deep learning of fMRI big data: a novel approach to subject-transfer decoding. arXiv preprint arXiv:1502.00093 (2015)

10. Kurmukov, A., et al.: Classifying phenotypes based on the community structure of human brain networks. In: Cardoso, M.J., et al. (eds.) GRAIL/MFCA/MICGen -2017. LNCS, vol. 10551, pp. 3–11. Springer, Cham (2017). https://doi.org/10. 1007/978-3-319-67675-3_1

11. Lynall, M.E., et al.: Functional connectivity and brain networks in schizophrenia. J. Neurosci. **30**(28), 9477–9487 (2010)

12. McKeown, M.J., Sejnowski, T.J.: Independent component analysis of fMRI data: examining the assumptions. Hum. Brain Mapp. **6**(5–6), 368–372 (1998)

13. Menon, V.: Large-scale brain networks and psychopathology: a unifying triple network model. Trends Cogn. Sci. **15**(10), 483–506 (2011)

14. Michel, V., Gramfort, A., Varoquaux, G., Eger, E., Keribin, C., Thirion, B.: A supervised clustering approach for fMRI-based inference of brain states. Pattern Recogn. **45**(6), 2041–2049 (2012)

15. Mitchell, T.M., et al.: Learning to decode cognitive states from brain images. Mach. Learn. **57**(1–2), 145–175 (2004)

16. Newman, S.D., Greco, J.A., Lee, D.: An fMRI study of the tower of London: a look at problem structure differences. Brain Res. **1286**, 123–132 (2009)

17. Onal, I., Ozay, M., Mizrak, E., Oztekin, I., Vural, F.T.Y.: A new representation of fMRI signal by a set of local meshes for brain decoding. IEEE Trans. Sig. Inf. Process. Netw. **3**(4), 683–694 (2017)

18. Onal, I., Ozay, M., Vural, F.T.Y.: Modeling voxel connectivity for brain decoding. In: 2015 International Workshop on Pattern Recognition in NeuroImaging (PRNI), pp. 5–8. IEEE (2015)

19. Ozay, M., Öztekin, I., Öztekin, U., Vural, F.T.Y.: Mesh learning for classifying cognitive processes. arXiv preprint arXiv:1205.2382 (2012)

20. Pereira, F., Mitchell, T., Botvinick, M.: Machine learning classifiers and fMRI: a tutorial overview. Neuroimage **45**(1), S199–S209 (2009)

21. Richiardi, J., Eryilmaz, H., Schwartz, S., Vuilleumier, P., Van De Ville, D.: Decoding brain states from fMRI connectivity graphs. Neuroimage **56**(2), 616–626 (2011)

22. Smith, S.M., Hyvärinen, A., Varoquaux, G., Miller, K.L., Beckmann, C.F.: Group-PCA for very large fMRI datasets. Neuroimage **101**, 738–749 (2014)

23. Tzourio-Mazoyer, N., et al.: Automated anatomical labeling of activations in SPM using a macroscopic anatomical parcellation of the MNI MRI single-subject brain. Neuroimage **15**(1), 273–289 (2002)

24. Vaidyanathan, P.: The theory of linear prediction. Synth. Lect. Sig. Process. **2**(1), 181–184 (2007)

25. Wang, X., Hutchinson, R., Mitchell, T.M.: Training fMRI classifiers to detect cognitive states across multiple human subjects. In: Advances in Neural Information Processing Systems, pp. 709–716 (2004)

26. Zhou, Z., Ding, M., Chen, Y., Wright, P., Lu, Z., Liu, Y.: Detecting directional influence in fMRI connectivity analysis using PCA based granger causality. Brain Res. **1289**, 22–29 (2009)

Proceedings of the First Workshop Beyond MIC: Integrating Imaging and Non-imaging Modalities for Healthcare Challenges

A Bayesian Disease Progression Model for Clinical Trajectories

Yingying Zhu$^{(\boxtimes)}$ and Mert R. Sabuncu$^{(\boxtimes)}$

Schools of ECE and BME, Cornell University, Ithaca, USA
{yz2377,ms3375}@cornell.edu

Abstract. In this work, we consider the problem of predicting the course of a progressive disease, such as cancer or Alzheimer's. Progressive diseases often start with mild symptoms that might precede a diagnosis, and each patient follows their own trajectory. Patient trajectories exhibit wild variability, which can be associated with many factors such as genotype, age, or sex. An additional layer of complexity is that, in real life, the amount and type of data available for each patient can differ significantly. For example, for one patient we might have no prior history, whereas for another patient we might have detailed clinical assessments obtained at multiple prior time-points. This paper presents a probabilistic model that can handle multiple modalities (including images and clinical assessments) and variable patient histories with irregular timings and missing entries, to predict clinical scores at future time-points. We use a sigmoidal function to model latent disease progression, which gives rise to clinical observations in our generative model. We implemented an approximate Bayesian inference strategy on the proposed model to estimate the parameters on data from a large population of subjects. Furthermore, the Bayesian framework enables the model to automatically fine-tune its predictions based on historical observations that might be available on the test subject. We applied our method to a longitudinal Alzheimer's disease dataset with more than 3,000 subjects [1] with comparisons against several benchmarks.

1 Introduction

Many progressive disorders, such as Alzheimer's disease (AD) [2], begin with mild symptoms that often precede diagnosis, and follow a patient-specific clinical trajectory that can be influenced by genetic and/or other factors. Therapeutic interventions, if available, are usually more effective in the earliest stages of a progressive disease. Therefore, tracking and predicting disease progression, particularly during the mild stages, is one of the primary objectives of personalized medicine.

In this paper, we are motivated by the real-world clinical setting where each individual is at risk and thus monitored for a specific progressive disease, such as AD. Furthermore, we assume that each individual might pay zero, one, or more

© Springer Nature Switzerland AG 2018
D. Stoyanov et al. (Eds.): GRAIL 2018/Beyond MIC 2018, LNCS 11044, pp. 57–65, 2018.
https://doi.org/10.1007/978-3-030-00689-1_6

visits to the clinic. In each clinical visit, various biomarkers or assessments (cor-related with the disease and/or its progression) are obtained. Example biomarker modalities include brain MRI scans, PET scans, blood tests, and cognitive test scores. The number and timing of the visits, and the exact types of data collected at each visit can be planned to be standardized, but often vary wildly between patients in practice. An ideal clinical prediction tool should be able to deal with this heterogeneity and compute accurate forecasts for arbitrary time horizons.

We present a probabilistic disease progression model that elegantly handles the aforementioned challenges of longitudinal clinical settings: data missingness, variable timing and number of visits, and multi-modal data (i.e., data of different types). The backbone of our model is a latent sigmoidal curve that captures the dynamics of the unobserved pathology, which is reflected in time-varying clini-cal assessments. Sigmoid curves are conceptually useful abstractions that fit well a wide range of dynamic physical and biological phenomena, including disease progression [3–5], which exhibit a floor and ceiling effect. In our framework, the sigmoid allows us to model the temporal correlation in longitudinal measure-ments and capture the dependence between the different tests and assessments, which are assumed to be generated conditionally independently from the latent state. We implemented an approximate Bayesian inference strategy on the pro-posed model and applied it to a large-scale longitudinal AD dataset [1].

In our experiments, we considered three target variables, which are widely used cognitive and clinical assessments associated with AD: the Mini Mental State Examination (MMSE) [6], the Alzheimer's Disease Assessment Scale Cog-nitive Subscale (ADAS-COG) [7], and the Clinical Dementia Rating Sum of Boxes (CDR-SB) [8]. We trained and evaluated the proposed model on a lon-gitudinal dataset with more than 3,000 subjects that included healthy controls (cognitively normal elderly individuals), subjects with mild cognitive impairment (MCI, a clinical stage that indicates high risk for dementia), and patients with AD. We provide a detailed analysis of prediction accuracy achieved with the proposed model and alternative benchmark methods under different scenarios that involve varying the past available visits and future time windows. In all our comparisons, the proposed model achieves significantly and substantially better accuracy for all target biomarkers.

2   Methods

2.1   Model

Let us first describe our notation and present our model. Assume we are given n subjects. $\mathbf{x}_i \in \mathbb{R}^{d \times 1}$ denotes subject i's d-dimensional attribute vector. In our experiments, this vector contains APOE genotype (encoded as number of E4 alleles, which can be $0, 1$ or 2) [9], education (in years) [10], sex (0 for female and 1 for male) [11] and two well-established neuroanatomical biomarkers of AD computed from a baseline MRI scan (namely total hippocampal [12] and ventricular volume [13] normalized by brain size). The MRI biomarkers capture so-called "brain reserve" [14]. Let $\mathbf{y}_i^k \in \mathbb{R}^{v_i \times 1}$ represent the values of the the

k'th dynamic (i.e., time-varying) target variable at v_i different clinical visits. $\mathbf{t}_i = [t_{i1}, \cdots, t_{iv_i}] \in \mathbb{R}^{v_i \times 1}$ denotes a vector of the age of subject i at these visits. The number and timing of the visits can vary across subjects. In general, we will assume $k \in \{1, \cdots, m\}$. In our experiments, we consider 3 target variables: MMSE, ADAS-COG or CDRSB and thus $m = 3$. We use $\mathbf{d}_i^k = [d_{i1}^k, \cdots, d_{iv_i}^k]$ to denote subject i's latent trajectory values associated with the k'th target variable. We assume each $d_{ij}^k \in [0, 1]$, with lower values corresponding to milder stages. As we describe below, the target variable, which is a clinical assessment, will be assumed to be a noisy observation of this latent variable. We model the latent trajectory of $\mathbf{d}_i^k$ as a sigmoid function of time (i.e., age), parameterized by a target- and subject-specific inflection point $p_i^k \in \mathbb{R}$ and a subject-specific slope parameter $s_i \in \mathbb{R}$. We assume that the slopes of the latent sigmoids associated with each target are coupled for each subject, yet the inflection points differ, which correspond to an average lag between the dynamics of target variables. This is consistent with the hypothesized biomarker trajectories of AD [3]. However, it would be easy to relax this assumption by allowing each target variable to have its own slope.

We assume the inflection points $\{p_i^k\}$ and slopes $\{s_i\}$ are random variables drawn from Gaussian priors with means equal to linear functions of subject-specific attributes $\mathbf{x}_i$: $p_i^k \sim \mathcal{N}(\mathbf{v}^T \mathbf{x}_i + a_k, \sigma_p^2), s_i \sim \mathcal{N}(\mathbf{w}^T \mathbf{x}_i + b, \sigma_s^2)$, where $a_k \in \mathbb{R}$ is associated with the k'th target (accounting for different time lags between target dynamics), while $\mathbf{v}, \mathbf{w} \in \mathbb{R}^{d \times 1}$, and $b, \sigma_p, \sigma_s \in \mathbb{R}$ are general parameters. Here and henceforth $\mathcal{N}(\mu, \sigma^2)$ denotes a Gaussian with mean μ and variance σ^2. Given s_i and p_i^k, the latent value d_{ij}^k associated with the k'th target is computed by evaluating the sigmoid at t_{ij}, $d_{ij}^k = \frac{1}{1 + \exp(-(t_{ij} - p_i^k) s_i)}$. The inflection point p_i^k marks the age at which the rate of change achieves its maximum, which is equal to $s_i/4$.

Finally, we assume that the target variable value y_{ij}^k is a linear function of the latent state d_{ij}^k corrupted by additive zero-mean independent Gaussian noise:

$$y_{ij}^k \sim \mathcal{N}(c_k d_{ij}^k + h_k, \sigma_k^2), \tag{1}$$

where c_k, h_k, and $\sigma_k \in \mathbb{R}$ are universal (not subject-specific) parameters associated with the k'th target variable. We refer to Eq. (1) as an observation model.

2.2   Inference

In this section, we discuss how to train the proposed model and apply it during test time.

Training: Let us use $\boldsymbol{\Theta}$ to denote the parameter set of our model:

$$\boldsymbol{\Theta} = \{\mathbf{w}, b, \sigma_p, \sigma_s, \mathbf{v}, \{a_k, c_k, h_k, \sigma_k\}_{k=1, \cdots, m}\}.$$

The goal of training is to estimate the model parameters $\boldsymbol{\Theta}$ given data from n subjects: $\{\mathbf{y}_i, \mathbf{x}_i, \mathbf{t}_i\}_{i=1,\ldots,n}$. Here, $\mathbf{y}_i = [\mathbf{y}_i^1 \ldots \mathbf{y}_i^m] \in \mathbb{R}^{v_i \times m}$ denotes m target values of the ith subject for v_i visits.

We estimate $\boldsymbol{\Theta}$ via maximizing the likelihood function:

$$\prod_{i=1}^{n} P(\mathbf{y}_i|\mathbf{x}_i, \mathbf{t}_i; \boldsymbol{\Theta}).$$

We use the standard notation of $p(y|x)$ to indicate the probability density function of the random variable Y (evaluated at y) conditioned on the random variable X taking on the value x. Also, parameters not treated as random variables are collected on the right hand side of ";".

Now, let us focus on the likelihood of each subject:

$$P(\mathbf{y}_i|\mathbf{x}_i, \mathbf{t}_i; \boldsymbol{\Theta}) = \int \int \left[\prod_{j=1}^{v_i} p(\mathbf{y}_{ij}|s_i, \mathbf{p}_i, \mathbf{t}_{ij}) \right] p(s_i, \mathbf{p}_i|\mathbf{x}_i; \boldsymbol{\Theta}) ds_i d\mathbf{p}_i,$$

with $p(s_i, \mathbf{p}_i|\mathbf{x}_i; \boldsymbol{\Theta})T = p(s_i|\mathbf{x}_i; \boldsymbol{\Theta})p(\mathbf{p}_i|\mathbf{x}_i; \boldsymbol{\Theta})T$.

Instead of the computationally challenging Eq. (2), we use variational approximation [15] and maximize the expected lower bound objective (ELBO):

$$F(\boldsymbol{\Theta}, \{\gamma_i\}) = \sum_{i=1}^{n} \mathbb{E}_q(\sum_{j=1}^{v_i} \sum_{k=1}^{m} T \log p(y_{ij}^k|s_i, p_i^k, \mathbf{t}_{ij}; \boldsymbol{\Theta}))$$

$$- \mathbb{E}_q(\log q(s_i; \gamma_i)) - \mathbb{E}_q(\log q(\mathbf{p}_i; \gamma_i)), \tag{2}$$

where $q(s_i; \gamma_i) = N(\mu_{si}, \sigma_{si}^2)$ and $q(\mathbf{p}_i; \gamma_i)) = N(\mu_{pi}, \boldsymbol{\Sigma}_{pi} = \boldsymbol{\Gamma}_{pi}^T \boldsymbol{\Gamma}_{pi})$ are proxy distributions that approximate the true posteriors $p(s_i|\mathbf{y}_i, \mathbf{x}_i; \boldsymbol{\Theta})$ and $p(\mathbf{p}_i|\mathbf{y}_i, \mathbf{x}_i; \boldsymbol{\Theta})$, respectively. During training, we use gradient-ascent to iteratively optimize Eq. 2 and solve for the optimal model parameters $\boldsymbol{\Theta}^*$ and the optimal parameters of the proxy distributions $\{\gamma_i^*\}$. The expectation in the first term is with respect to the proxy distributions and can be approximated via Monte Carlo sampling:

$$\mathbb{E}_q(\sum_k \sum_j \log p(y_{ij}^k|s_i, p_i^k, \mathbf{t}_{ij}; \boldsymbol{\Theta})) \approx \frac{1}{S} \sum_j \sum_{s=1}^{S} \log p(\mathbf{y}_{ij}|s_i^{(s)}, \mathbf{p}_i^{(s)}, \mathbf{t}_{ij}; \boldsymbol{\Theta}), \tag{3}$$

where $s_i^{(s)}$ and $\mathbf{p}_i^{(s)}$ are samples drawn using the "re-parameterization trick." I.e., $s_i^{(s)} = \eta^{(s)} \sigma_{si} + \mu_{si}$ and $\mathbf{p}_i^{(s)} = \boldsymbol{\Gamma}_{pi}^T \epsilon^{(s)} + \mu_{pi}$, where $\eta^{(s)} \in \mathbf{R}$ and $\epsilon^{(s)} \in \mathbf{R}^{m \times 1}$ are realizations of the auxiliary random variables, independently drawn from zero-mean standard Gaussians, $N(0, 1)$ and $N(\mathbf{0}, \mathbf{I})$, respectively. The "re-parameterization trick" allows us to differentiate the ELBO (or more accurately, its approximation that uses Eq. 3) with respect to γ_i.

E.g.:

$$\frac{\partial s_i^{(s)}}{\partial \sigma_{si}} = \eta^{(s)}, \text{ and } \frac{\partial s_i^{(s)}}{\partial \mu_{si}} = 1.$$

Testing. During test time, we are interested in computing the posterior distribution of $\mathbf{y}_{n+1}$ for a new subject with $\mathbf{x}_{n+1}$ at an arbitrary time-point (age) t.

We drop the second sub-script, i.e., j index, of $\mathbf{y}_{n+1}$ to emphasize that we will be computing these posterior probabilities at many different (often future) time-points. There are two types of test subjects: those with no history of visits (scenario 1), and those with at least one prior clinical visit (scenario 2). For scenario 2, we will use $\{\mathbf{y}_{(n+1)j}, t_{(n+1)j}\}_{j=1,\dots,v_{n+1}}$ to collectively denote the v_{n+1} historical observations and their corresponding visit times. We fix $\boldsymbol{\Theta}^*$ to the values obtained from training. In scenario 1, we use Eq. (eq:ELBO) to compute the posterior. In the second scenario, we will first maximize the ELBO of Eq. (2) with respect to γ_{n+1} and evaluated for the observations on the new subject $\{\mathbf{y}_{(n+1)j}, t_{(n+1)j}\}$ and attribute vector: $\mathbf{x}_{n+1}$. We then proceed to use these approximate q distributions in Eq. (2), replacing $p(s|\mathbf{x}_i; \boldsymbol{\Theta}^*)$ and $p(p^k|x_i; \boldsymbol{\Theta}^*)$, to evaluate the posterior distribution for an arbitrary time-point t conditioned on past observations.

3 Experiments

Dataset. We use a dataset of 3,057 subjects (baseline age 73.3 ± 17.2 years) collected by ADNI [1] to empirically validate and demonstrate the proposed model. This dataset contained multiple clinical visits per subject, during which thorough cognitive and symptomatic assessments were conducted. In our experiments, we used MMSE, ADAS-COG and CDR-SB as three target variables. MMSE has a range between 0 (impaired) and 30 (healthy), whereas ADAS-COG takes on values between 0 (healthy) to 70 (severe), and CDR-SB varies from 0 (healthy) to 18 (severe). The first two (MMSE and ADAS-COG) are general cognitive assessments that track and predict dementia, while CDR-SB is a clinical score that measures the severity of dementia-associated symptoms. In addition to the target variables, we utilized individual-level traits associated with AD: age, APOE genotype (number of E4 alleles), sex, and education (in years). We also used baseline brain MRI scans to derive two anatomical biomarkers of AD: total hippocampal and ventricle volume normalized by brain size. These imaging biomarkers were automatically computed with FreeSurfer [16] and quality controlled as previously described [17].

3.1 Experimental Setup

Benchmark Methods. In our experiments, we compare the proposed method to the following benchmarks:

1. **Global:** A 4-parameter (scale, bias, inflection, and slope) sigmoidal model that was fit on all training data (least-squares).
2. **Sex-specific:** Same as "Global" but separate for males and females.
3. **APOE-specific:** Same as "Global", but separate for three groups defined by APOE-E4 allele count $\{0, 1, 2\}$.
4. **Sex- and APOE-specific:** Same as "Global", but separate for each sex and APOE group.

5. **Linear mixed effects (LME) model:** A linear regression model with subject-specific attributes ($\mathbf{x}_i$) as fixed effects, and time and bias term as a random effects. This LME model, commonly used to capture longitudinal dynamics, allows each subject to deviate from the average trajectory determined by its attributes by shifts in slope and offset.

6. **Subject-specific linear model:** Least-squares fit of a linear model on each subject's historical data. When there is only one past visit, we adopt a carry-forward extrapolation.

Implementation of Proposed Method. We coded in Python [1], using the Edward library [18], which is in turn built on TensorFlow [19]. We used a 20-fold cross-validation strategy in all our experiments. We first partitioned the data into 20 non-overlapping, roughly equally-sized sets of subjects. In each of the 20 folds, we reserved one of the partitions as the independent test set. Out of the remaining 19 partitions, one was set aside as a validation set, while the rest were combined into a training set. The training set was used to estimate the model parameters, i.e., Θ^*, while performance on the validation set was used to select hyper-parameters, such as step size in the optimization and evaluate random initializations. Finally, test performance was computed on the test set. We report results averaged across 20 folds.

3.2 Results and Discussion

We first show quantitative prediction results for all methods and target variables (MMSE, ADAS-COG, and CDRSB). In the following, we consider several prediction scenarios. In the first scenario, we vary the number of past visits available on test subjects (i.e., v_{n+1}). In general, we expect this variation to influence the LME and subject-specific linear model benchmarks, in addition to the proposed model. These methods fine-tune their predictions based on historical observations available on test data. With more test observations, we expect them to achieve better accuracy. All other benchmarks are fixed after training and thus their performance should not improve with increasing number of past observations. In the second scenario, we fix the number of past observations on test subjects and vary the prediction horizon. In general, all models' predictions should be less accurate for more distant future time-points.

Varying the Number of Past Visits. Figure 1 shows the MMSE, ADAS-COG and CDRSB prediction accuracies (mean and standard deviation of absolute error). We observe that the performance of the training-fixed benchmarks (1–4) worsen slightly as the number of past visits increases. This is likely because the training data contains more samples at early times (i.e., relatively younger ages), partially because most subjects drop out by their 4th visit. Therefore, a model trained on these data is expected to be less accurate for older ages.

The adaptive benchmarks (5–6) and the proposed model, on the other hand, overcome this handicap to achieve better accuracy with more past visits. As we

---

[1] The code of this work is available at https://github.com/zyy123jy/kdd.

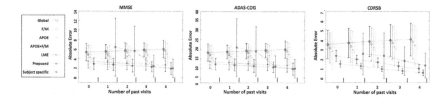

Fig. 1. Absolute error (mean and standard derivation) of all methods for predicting MMSE, ADAS-COG and CDRSB, as a function of number of past visits available on test subjects.

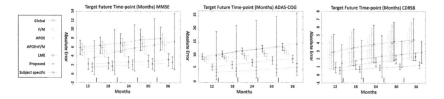

Fig. 2. Absolute error (mean and standard derivation) of all methods for predicting MMSE, ADAS-COG and CDRSB. We used two points from each test subject as past observations and varied the time horizon for prediction.

discussed above, this is largely because these techniques exploit test observations to fine-tune their models. The subject-level linear model (benchmark 6), in fact, is an extreme example, where the predictions are computed merely by extrapolating from historical observations without relying on training data.

Finally, the proposed model achieves a significantly and substantially better accuracy than all benchmarks (all paired permutation p-values $< p_{max} = 0.04$). The subject-specific benchmark (6) exhibits the largest variance implying the quality of performance varies wildly across subjects. Overall, the training-fixed benchmarks perform the worst. In general the proposed model's variance is among the smallest, indicating consistency in prediction accuracy.

Varying the Time Horizon. In order to evaluate how prediction performance changes as a function of the time horizon, we evaluated the methods for different future time-points. In this empirical scenario, we assume that each test subject has 2 past clinical assessments (obtained at baseline and month 6). Our goal is to predict MMSE, ADAS-COG and CDRSB scores at later time-points (starting at 12 months after baseline, up to 36 months). Based on the longitudinal study protocol, we considered 6 month intervals and assigned the actual visits to the closest 6-month bucket.

Figure 2 shows prediction accuracies of all considered methods. The proposed method performs significantly (all paired permutation p-values $< p_{max} = 0.03$) and substantially better than all other methods, with the difference increasing from the short term (12 months) to long term (36 months). For the benchmark models, prediction accuracy tends to drop more dramatically for longer time horizons. As above, training-fixed benchmarks perform the worst.

4   Conclusion

We presented a probabilistic, latent disease progression model for capturing the dynamics of the underlying pathology that is often shaped by risk factors such as genotype. Our work was motivated by real-world clinical applications, where irregular visiting patterns, missing variables, and inconsistent multi-modal assessments are ubiquitous. We applied the proposed method on a large dataset of Alzheimer's disease for predicting clinical scores at varying time horizons with promising results. Future work will conduct a more detailed analysis of our proposed model. We are also interested in exploring the use of modern neural network based methods, such as Recurrent Neural Networks [20], for this application.

Acknowledgements. This work was supported by NIH grants R01LM012719, R01AG053949, and 1R21AG050122, and the NSF NeuroNex grant 1707312. We used data from Tadpole 2017 Challenge (https://tadpole.grand-challenge.org/home/).

References

1. Petersen, R.: Alzheimer's Disease Neuroimaging Initiative (ADNI): clinical characterization. Neurology **74**, 201–209 (2010)
2. Alzheimer's Association: Alzheimer's disease facts and figures. Alzheimer's Dement. 158–194 (2010)
3. Jack, C., et al.: Hypothetical model of dynamic biomarkers of the Alzheimer's pathological cascade. Lancet Neurol. **9**, 119–128 (2010)
4. Dalca, A.V., Sridharan, R., Sabuncu, M.R., Golland, P.: Predictive modeling of anatomy with genetic and clinical data. In: Navab, N., Hornegger, J., Wells, W.M., Frangi, A.F. (eds.) MICCAI 2015 Part III. LNCS, vol. 9351, pp. 519–526. Springer, Cham (2015). https://doi.org/10.1007/978-3-319-24574-4_62
5. Haxby, J., Raffaele, K., Gillette, J., Schapiro, M., Rapoport, S.: Individual trajectories of cognitive decline in patients with dementia of the Alzheimer type. J. Clin. Exp. Neuropsychol. **14**(4), 575–592 (1992)
6. Marta, M.: Modelling mini mental state examination changes in Alzheimer's disease (2000)
7. Cano, S.J., et al.: The ADAS-COG in Alzheimer's disease clinical trials: psychometric evaluation of the sum and its parts. J. Neurol. Neurosurg. Psychiat. **81**, 1363–1368 (2010)
8. O'Bryant, S., et al.: Staging dementia using clinical dementia rating scale sum of boxes scores: a Texas Alzheimer's research consortium study. Archiv. Neurol. **65**, 1091–1095 (2008)
9. Corder, E.H., et al.: Gene dose of Apolipoprotein E type 4 allele and the risk of Alzheimer's disease in late onset families. Science **261**(5123), 921–923 (1993)
10. Katzman, R.: Education and the prevalence of dementia and Alzheimer's disease. Neurology **43**, 13–20 (1993)
11. Fratiglioni, L., et al.: Prevalence of Alzheimer's disease and other dementias in an elderly urban population relationship with age, sex, and education. Neurology **41**, 1886–1892 (1991)

12. Jack, C., et al.: Prediction of AD with MRI-based hippocampal volume in mild cognitive impairment. Neurology **52**, 1397–1403 (1999)
13. Nestor, S., et al.: Ventricular enlargement as a possible measure of Alzheimer's disease progression validated using the Alzheimer's disease neuroimaging initiative database. Brain **131**, 2443–2454 (2008)
14. Stern, Y.: Cognitive reserve in ageing and Alzheimer's disease. Lancet Neurol. **11**(11), 1006–1012 (2012)
15. Ranganath, R., et al.: Black box variational inference. In: Artificial Intelligence and Statistics (2014)
16. Fischl, B.: FreeSurfer. Neuroimage **62**, 774–781 (2012)
17. Mormino, E., et al.: Polygenic risk of Alzheimer disease is associated with early-and late-life processes. Neurology **87**, 481–488 (2016)
18. Tran, D., et al.: Edward: a library for probabilistic modeling, inference, and criticism. arXiv preprint arXiv:1610.09787 (2016)
19. Abadi, M., et al.: TensorFlow: a system for large-scale machine learning. In: OSDI (2016)
20. Che, Z., Purushotham, S., Cho, K., Sontag, D.A., Liu, Y.: Recurrent neural networks for multivariate time series with missing values. Sci. Rep. **8** (2018)

Multi-modal Brain Connectivity Study Using Deep Collaborative Learning

Wenxing Hu[1], Biao Cai[1], Vince Calhoun[2], and Yu-Ping Wang[1(✉)]

[1] Biomedical Engineering Department, Tulane University,
New Orleans, LA 70118, USA
wyp@tulane.edu
[2] The Mind Research Network and Department of ECE, University of New Mexico,
Albuquerque, NM 87106, USA
vcalhoun@mrn.org

Abstract. Functional connectivities in the brain explain how different brain regions interact with each other when conducting a specific activity. Canonical correlation analysis (CCA) based models, have been used to detect correlations and to analyze brain connectivities which further help explore how the brain works. However, the data representation of CCA lacks label related information and may be limited when applied to functional connectivity study. Collaborative regression was proposed to address the limitation of CCA by combining correlation analysis and regression. However, both prediction and correlation are sacrificed as linear collaborative regression use the same set of projections on both correlation and regression. We propose a novel method, deep collaborative learning (DCL), to address the limitations of CCA and collaborative regression. DCL improves collaborative regression by combining correlation analysis and label information using deep networks, which may lead to better performance both for classification/prediction and for correlation detection. Results demonstrated the out-performance of DCL over other conventional models in terms of classification accuracy. Experiments showed the difference of brain connectivities between different age groups may be more significant than that between different cognition groups.

Keywords: Canonical correlation · Deep network · fMRI
Functional connectivity

1 Introduction

Brain connectivity depicts the functional relations between different brain regions [1]. Investigating time-varying dynamic changes in brain connectivity has been increasingly studied in recent years [2]. Many works [3–5] have studied brain connectivity and investigated how brain connectivity changes during adolescence and how it differs between different age groups, e.g., children and young adults.

© Springer Nature Switzerland AG 2018
D. Stoyanov et al. (Eds.): GRAIL 2018/Beyond MIC 2018, LNCS 11044, pp. 66–73, 2018.
https://doi.org/10.1007/978-3-030-00689-1_7

A number of statistical learning models, e.g. group independent component analysis [6] and canonical correlation analysis (CCA) [7], have been applied to multi-modal study to analyze complimentary information between different imaging modalities and also applied to imaging-genetic study to detect interactions between genetic factors [8], e.g. single nucleotide polymorphisms (SNP), and endo-phenotypes, e.g., functional magnetic resonance imaging (fMRI). Among these methods, CCA has been widely used to detect multivariate correlations between two datasets. CCA reduces data dimensionality by projecting higher dimensional data into lower dimensional spaces. Many variants of CCA, e.g., multiple CCA [9], multi-set CCA [10], sparse CCA [11], structured sparse CCA [12], have been developed to address more specific challenges in real data applications. Despite the wide application of CCA, canonical variables lack label related information, which may be a limitation to CCA's application and restrict the interpretation of its output. To address the limitation, Gross et al. [13] proposed a model, collaborative regression, which identifies label related correlations by incorporating regression into CCA's objective function. However, according to the simulation results in [13], collaborative regression may result in poor performance for prediction. This may be due to the restriction on coefficient vectors which requires the projection of correlation and that of the regression to be in the same direction.

In this paper, we proposed a novel model, deep collaborative learning (DCL), which addresses the limitation of collaborative regression by combining correlation analysis and regression method via deep networks which may lead to higher classification accuracies and better correlation detection. The performance of DCL model was verified by the experiments in our work. In addition, many interesting discoveries about brain connectivity were found.

The rest of the paper is organized as follows. The limitation of existing methods and how the proposed model addresses the limitations were introduced in Sect. 2. Section 3 introduces the collection and preprocessing of brain connectivity data. Conclusions and discussion of the results and possible limitations/extensions of the work were in Sect. 4.

2   Method

2.1   Overview of Linear Canonical Correlation Analysis (CCA)

Canonical correlation analysis (CCA) [7] is a model widely used for analyzing linear correlations between two data. It projects original data into the optimal directions (canonical loading vectors) with the highest Pearson correlation.

Suppose we have two data matrices $X_1 \in \mathbb{R}^{n \times p}, X_2 \in \mathbb{R}^{n \times q}$, CCA seeks two projection matrices U_1 and U_2 by optimizing the following objective function

$$(U_1^*, U_2^*) = \underset{U_1, U_2}{\operatorname{argmax}} \operatorname{Trace}(U_1' \Sigma_{12} U_2) \tag{1}$$

subject to $U_1' \Sigma_{11} U_1 = U_2' \Sigma_{22} U_2 = \mathbf{I}_n$; where $U_1 \in \mathbb{R}^{p \times k}$,

$U_2 \in \mathbb{R}^{q \times k}, k = \min(\operatorname{rank}(X_1), \operatorname{rank}(X_2)), \Sigma_{ij} := X_i' X_j$

2.2   Deep CCA

Deep CCA was proposed by Andrew et al. [14] to detect nonlinear cross-data correlations. As illustrated in Fig. 1(a), deep CCA introduces a deep network representation before applying CCA framework. Unlike linear CCA, which seeks the optimal loading matrices U_1, U_2, deep CCA seeks the optimal network representation $f_1(X_1)$, $f_2(X_2)$, as shown in Eq. (2).

$$(f_1^*, f_2^*) = \underset{f_1, f_2}{\mathrm{argmax}} \left\{ \max_{U_1, U_2} \frac{U_1' f_1'(X_1) f_2(X_2) U_2}{\|f_1(X_1)U_1\|_2 \|f_2(X_2)U_2\|_2} \right\} \tag{2}$$

where f_1, f_2 are two deep networks as illustrated in Fig. 1(a).

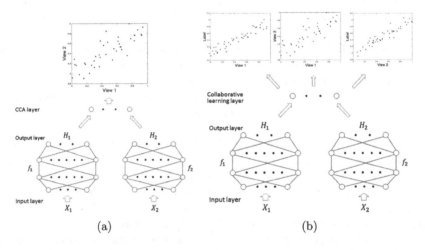

(a)   (b)

Fig. 1. A figure showing the work-flows of deep CCA and deep collaborative learning. Data X_1, X_2 are input; deep networks f_1, f_2 work on X_1, X_2 and yield H_1, H_2 as output, to which CCA is or collaborative regression was applied subsequently. For deep CCA, the optimization problem is to find the optimal network $\hat{f}_1$, $\hat{f}_2$ with the highest canonical correlation. For deep collaborative learning, the optimization problem is to find the optimal network $\hat{f}_1$, $\hat{f}_2$ which give both the highest canonical correlation and the smallest prediction error

The introduction of deep network representation leads to a more flexible ability to detect both linear and nonlinear correlations. According to experiments on both speech data and handwritten digits data [14], deep CCA's representation was more correlated than that by other correlation analysis methods, e.g., linear CCA, kernel CCA.

2.3   Deep Collaborative Learning (DCL)

CCA, as well as deep CCA, is a method of data representation. However, CCA based methods have not found wide application compared with PCA based methods. As a method of dimension reduction, CCA's output (canonical variables)

lacks connections to label information and the detected correlations may be difficult to interpret consequently. To address the limitation of CCA, Gross et al. [13] proposed a new model, called collaborative regression, whose formulation is shown in (3). Specifically, given a label data $Y \in \mathbb{R}^{n \times 1}$, collaborative regression maximizes the following objective function

$$(u_1^*, u_2^*) = \underset{u_1, u_2}{\operatorname{argmax}} b_1 \|X_2 u_2 - X_1 u_1\|_2 \tag{3}$$
$$+ b_2 \|Y - X_1 u_1\|_2 + b_3 \|Y - X_2 u_2\|_2$$

Collaborative regression addresses CCA's limitations by taking advantage of label information so that it can detect canonical correlations which are label related. However, according to the simulation in [13], collaborative regression may lead to poor performance in terms of classification accuracies and therefore may not be suitable for brain connectivity study. This may be due to the coupled restriction on coefficient vectors u_1, u_2 which requires the projection of correlation and that of the regression to be in the same direction.

To address these limitations of both CCA and collaborative regression method, we propose a novel model, deep collaborative learning (DCL), which incorporates regression into CCA in an uncoupled way via deep networks. Suppose we have two modality data $X_1 \in \mathbb{R}^{n \times p}, X_2 \in \mathbb{R}^{n \times q}$ and a label data $Y \in \mathbb{R}^{n \times 1}$, where n denotes sample size (number of subjects) and p, q are the dimensionality of feature of X_1, X_2 respectively. The formulation of deep collaborative learning is shown in Eqs. (4) and (5) and its framework is illustrated in Fig. 1(b).

$$(H_1^*, H_2^*) = \underset{H_1, H_2}{\operatorname{argmax}} \{ \underset{U_1, U_2}{\max} \operatorname{Trace}(U_1' H_1' H_2 U_2) + \underset{\beta_1}{\max} \operatorname{Trace}(\beta_1' H_1' Y) \|Y\|_2^{-1} \tag{4}$$
$$+ \underset{\beta_2}{\max} \operatorname{Trace}(\beta_2' H_2' Y) \|Y\|_2^{-1} \}$$
$$= \underset{H_1, H_2}{\operatorname{argmax}} \{ \| \Sigma_{11}^{-\frac{1}{2}} \Sigma_{12} \Sigma_{22}^{-\frac{1}{2}} \|_{tr} + \operatorname{Trace}(\Sigma_{11}^{-\frac{1}{2}} H_1' Y) \|Y\|_2^{-1} \tag{5}$$
$$+ \operatorname{Trace}(\Sigma_{22}^{-\frac{1}{2}} H_2' Y) \|Y\|_2^{-1} \}$$

where $H_1 = f_1(X_1) \in \mathbb{R}^{n \times r}$, $H_2 = f_2(X_2) \in \mathbb{R}^{n \times s}$; f_1, f_2 are two deep networks as illustrated in Fig. 1(b); $\Sigma_{ij} := H_i' H_j$; and $\|A\|_{tr} := \operatorname{Trace}(\sqrt{A'A}) = \Sigma \sigma_i$; U_1, U_2 in Eq. (4) subject to $U_1' \Sigma_{11} U_1 = U_2' \Sigma_{22} U_2 = \mathbf{I}$.

As shown in Eqs. (4) and (5), deep collaborative learning seeks the optimal network representation $H_1 = f_1(X_1), H_2 = f_2(X_2)$ instead of the optimal projection vectors $u_1, u_2, \beta_1, \beta_2$ and the coupled restriction can be relaxed consequently. Relaxation of the coupled restriction leads to a better performance on both prediction/classification and correlation analysis compared with linear collaborative regression.

3   Application to Brain Connectivity Study

3.1   Introduction of Brain Connectivity

We next apply the DCL model to the study of brain connectivity and development. Brain connectivity depicts the anatomical or functional associations between different brain regions or nodes [1]. It is of interest to investigate how brain connectivity changes during adolescence and how it differs between different age groups, e.g., children, young adults, which may further contributes to the study of normal and pathological brain development. The proposed model, deep collaborative learning, is a network representation based model which can detect signals having both strong correlations (reflecting brain connectivity) and good discriminative power (reflecting differences between age groups) and therefore is very suitable for the study of brain connectivity and development.

3.2   Brain Connectivity Data

Several brain fMRI modalities from the Philadelphia Neurodevelopmental Cohort (PNC) [15] were used in the experiments. PNC cohort is a large-scale collaborative study between the Brain Behavior Laboratory at the University of Pennsylvania and the Children's Hospital of Philadelphia. It contains multimodal neuroimaging data (e.g., fMRI, diffusion tensor imaging) and multiple genetic factors (e.g., singular nucleotide polymorphisms of SNPs) from adolescents aged from 8 to 21 years. There were three types of fMRI data in PNC cohort which were collected during different task states: resting-state fMRI (rs-fMRI), emotion task fMRI (emoid t-fMRI), and nback task fMRI (nback t-fMRI). Two types of labels, age and Wide Range Achievement Test (WRAT) score [16], which is a measure of comprehensive cognitive ability, were used for classification and correlation analysis.

3.3   Results

We compared the performance of the DCL model to that of CCA, deep CCA (DCCA), collaborative regression (CR) for both age classification and the classification of cognitive ability. For age groups, the top 20% (in terms of age) subjects were extracted as young adults group (aged 18 to 22) while the bottom 20% were extracted as children group (aged 8 to 11). For cognitive ability group, the top 20% (assess via the WRAT score) of individuals were extracted as a high cognition group (WRAT 114–145) while the bottom 20% were extracted as a low cognition group (WRAT 55–89). Data were separated into a training set (60%) and a testing set (40%). The training set was used for DCL's network training and the trained network was applied to testing set for classification subsequently. All preprocessing methods, including data augmentation, data standardization, etc., were performed on training set and testing set separately. All hyper-parameters, including momentum, activation function type, learning rate, decay rate, batch size, maximum epochs, the number of layers,

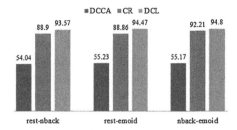

Fig. 2. A figure showing the comparison of the performances of different methods on classifying different age groups (young adults (aged 18–22) vs. children (aged 8–11)). The full names of the methods are deep CCA (DCCA), collaborative regression (CR), deep collaborative learning (DCL). The numbers appearing in the figure were classification accuracies (%).

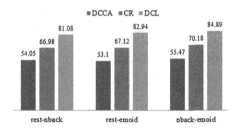

Fig. 3. A figure showing the comparison of the performances of different methods on classifying high/low WRAT scores (cognitive ability). The full names of the methods can be found in the caption of Fig. 2. The numbers appearing in the figure were classification accuracies (%).

the number of nodes in each layer, and the dimensionality of canonical variables, were chosen using grid search based on the training data. To verify the performance of the DCL model, we also included the results of other competitive methods, including deep CCA and collaborative regression (CR). As CCA based methods require at least two datasets as input, different data-pair combinations were used as data input: rs-fMRI and nback t-fMRI (rest-nback); rs-fMRI and emoid t-fMRI (rest-emoid); nback t-fMRI and emoid t-fMRI (rest-emoid). For each data combination, we tested the performance of deep CCA, CR, and DCL, and the results were shown in Fig. 2 (classifying age groups) and Fig. 3 (classifying WRAT groups). We only included accuracy as a criterion for evaluating classification performance as the two groups had balanced numbers of subjects (top 20% versus bottom 20%).

From Figs. 2 and 3, the proposed model, deep collaborative learning, achieved higher classification accuracies than two CCA based models and collaborative regression for both classifying age groups and classifying cognition groups, which may be a result of the nonlinear representation of deep network and the combination of prediction and correlation detection. Collaborative regression performed better than deep CCA but worse than DCL in terms of classification, which may be due to the incorporation of label information. The high classification accuracy (over 90%) indicates that different age groups (e.g. young adults and children) and different cognition groups (high WRAT scores and low WRAT scores) may exhibit different brain functional connectivity patterns and functional brain connectivity might be used as a finger-print to identify different subjects. In addition, it can also be seen from Figs. 2 and 3 that the classification accuracy of age groups is higher than that of cognition groups which might be due to the fact that age is a fixed phenotype while cognition score is just a rough measure which is not as accurate and consistent as age.

4   Discussion and Conclusion

In the work we propose a new model, DCL, which captures label related correlations and performs well on classification by combining correlation analysis and regression using deep networks. According to the results, DCL performed better than deep CCA and collaborative regression, which may demonstrate that the relaxation of restriction on projections using deep networks help achieve higher classification accuracies. The superior power of DCL on both correlation detection and classification makes DCL a suitable model for brain connectivity study, whose research interest focuses on analyzing correlations of functional networks and how different subject groups exhibit different brain connectivity patterns. From the results, both different age groups and different cognition groups exhibit significant differences in brain connectivities. In addition, brain connectivity tends to be more discriminative when used to classify age groups than to classify WRAT/cognition groups. The framework of DCL can be easily extended to more than three datasets integration as in [17] and may become more suitable to deal with brain imaging data if replacing fully connected networks with convolutional neural networks.

Acknowledgment. The authors would like to thank the NIH (R01 GM109068, R01 MH104680, R01 MH107354, P20 GM103472, R01 REB020407, R01 EB006841) and NSF (#1539067) for the partial support.

References

1. Calhoun, V.D., Adali, T.: Time-varying brain connectivity in fMRI data: whole-brain data-driven approaches for capturing and characterizing dynamic states. IEEE Sig. Process. Mag. **33**(3), 52–66 (2016)
2. Calhoun, V.D., Miller, R., Pearlson, G., Adalı, T.: The chronnectome: time-varying connectivity networks as the next frontier in fMRI data discovery. Neuron **84**(2), 262–274 (2014)
3. Braun, U., Muldoon, S.F., Bassett, D.S.: On human brain networks in health and disease. eLS, 1–9 (2001)
4. Cai, B., Zille, P., Stephen, J.M., Wilson, T.W., Calhoun, V.D., Wang, Y.P.: Estimation of dynamic sparse connectivity patterns from resting state fMRI. IEEE Trans. Med. Imaging **37**(5), 1224–1234 (2018)

5. Deng, S.-P., Hu, W., Calhoun, V.D., Wang, Y.-P.: Integrating imaging genomic data in the quest for biomarkers for schizophrenia disease. IEEE/ACM Trans. Comput. Biol. Bioinform. (2017)

6. Calhoun, V.D., Liu, J., Adalı, T.: A review of group ICA for fMRI data and ICA for joint inference of imaging, genetic, and ERP data. Neuroimage **45**(1), S163–S172 (2009)

7. Hotelling, H.: Relations between two sets of variates. Biometrika **28**(3/4), 321–377 (1936)

8. Cao, S., Qin, H., Gossmann, A., Deng, H.-W., Wang, Y.-P.: Unified tests for fine-scale mapping and identifying sparse high-dimensional sequence associations. Bioinformatics **32**(3), 330–337 (2015)

9. Hu, W., et al.: Adaptive sparse multiple canonical correlation analysis with application to imaging (epi) genomics study of schizophrenia. IEEE Trans. Biomed. Eng. **65**(2), 390–399 (2018)

10. Li, Y.-O., Adali, T., Wang, W., Calhoun, V.D.: Joint blind source separation by multiset canonical correlation analysis. IEEE Trans. Sig. Process. **57**(10), 3918–3929 (2009)

11. Lin, D., Calhoun, V.D., Wang, Y.-P.: Correspondence between fMRI and SNP data by group sparse canonical correlation analysis. Med. Image Anal. **18**(6), 891–902 (2014)

12. Du, L., et al.: Structured sparse canonical correlation analysis for brain imaging genetics: an improved graphnet method. Bioinformatics **32**(10), 1544–1551 (2016)

13. Gross, S.M., Tibshirani, R.: Collaborative regression. Biostatistics **16**(2), 326–338 (2014)

14. Andrew, G., Arora, R., Bilmes, J., Livescu, K.: Deep canonical correlation analysis. In: International Conference on Machine Learning, pp. 1247–1255 (2013)

15. Satterthwaite, T.D., et al.: Neuroimaging of the philadelphia neurodevelopmental cohort. Neuroimage **86**, 544–553 (2014)

16. Wilkinson, G.S., Robertson, G.J.: Wide Range Achievement Test. Psychological Assessment Resources (2006)

17. Hu, W., Lin, D., Calhoun, V.D., Wang, Y.-P.: Integration of SNPs-FMRI-methylation data with sparse multi-CCA for schizophrenia study. In: 2016 IEEE 38th Annual International Conference of the Engineering in Medicine and Biology Society (EMBC), pp. 3310–3313. IEEE (2016)

Towards Subject and Diagnostic Identifiability in the Alzheimer's Disease Spectrum Based on Functional Connectomes

Diana O. Svaldi[1]($\boxtimes$) (iD), Joaquín Goñi[1,2]($\boxtimes$) (iD),
Apoorva Bharthur Sanjay[1], Enrico Amico[2], Shannon L. Risacher[1],
John D. West[1], Mario Dzemidzic[1], Andrew Saykin[1],
and Liana Apostolova[1]

[1] Indiana University School of Medicine, Indianapolis, IN 46202, USA
dosvaldi@iu.edu, jgonicor@purdue.edu
[2] Purdue University, Lafayette, IN 47907, USA

Abstract. Alzheimer's disease (AD) is the only major cause of mortality in the world without an effective disease modifying treatment. Evidence supporting the so called "disconnection hypothesis" suggests that functional connectivity biomarkers may have clinical potential for early detection of AD. However, known issues with low test-retest reliability and signal to noise in functional connectivity may prevent accuracy and subsequent predictive capacity. We validate the utility of a novel principal component based diagnostic identifiability framework to increase separation in functional connectivity across the Alzheimer's spectrum by identifying and reconstructing FC using only AD sensitive components or connectivity modes. We show that this framework (1) increases test-retest correspondence and (2) allows for better separation, in functional connectivity, of diagnostic groups both at the whole brain and individual resting state network level. Finally, we evaluate a posteriori the association between connectivity mode weights with longitudinal neurocognitive outcomes.

Keywords: Alzheimer's disease · Functional connectivity
Principal component analysis · Resting state fMRI

1 Introduction

Developing biomarkers for early detection of Alzheimer's disease (AD) is of critical importance as researchers believe clinical trial failures are in part due to testing of therapeutic agents too late in the disease [1]. The AD disconnection syndrome hypothesis [2] posits that AD spreads via propagation of dysfunctional signaling, indicating that functional connectivity (FC) biomarkers have potential for early detection. Despite this potential, known issues with high amounts of variability in acquisition and preprocessing of resting state fMRI, and ultimately low disease-related signal to noise ratio in FC [3], remain a critical barrier to incorporating FC as a clinical biomarker of AD. Recent work validated the utility of group level principal component analysis (PCA) to denoise FC by reconstructing subject level FC using PCs which

© Springer Nature Switzerland AG 2018
D. Stoyanov et al. (Eds.): GRAIL 2018/Beyond MIC 2018, LNCS 11044, pp. 74–82, 2018.
https://doi.org/10.1007/978-3-030-00689-1_8

optimized test-retest reliability through a measurement denominated differential identifiability [4]. Building on this work, we expand the utility of the framework to increase separation across diagnostic groups in the AD spectrum by reconstructing individual FC using AD sensitive PCs. We identify AD sensitive PCs using a novel diagnostic identifiability metric (D). We evaluate the proposed method with data from the Alzheimer's Disease Neuroimaging Initiative (ADNI2/GO) using group balanced, bootstrapped random sampling.

2 Methods

2.1 Subject Demographics

Of the original 200 ADNI2/GO individuals with resting state fMRI scans, subjects were excluded if they (1) had only extended resting state scans, (2) had no Amyloid status provided, (3) were cognitively impaired, but Amyloid-beta protein negative (Aβ−) negative, and/or had (4) over 30% of fMRI time points censored (see Sect. 2.2). The final sample included 82 individuals. Only Aβ positive (Aβ+) individuals were included in cognitively impaired groups to avoid confounding by non-AD neurodegenerative pathologies. Subjects were sorted into 5 diagnostic groups using criterion from ADNI2/GO and Aβ positivity: (1) normal controls (CN$_{Aβ-}$, n = 15), (2) pre-clinical AD (CN$_{Aβ+}$, n = 12), (3) early mild cognitive impairment (EMCI$_{Aβ+}$, n = 22), (4) late mild cognitive impairment (LMCI$_{Aβ+}$, n = 12), and (5) dementia (AD$_{Aβ+}$, n = 21). Aβ status was determined using either mean PET standard uptake value ratio cutoff (Florbetapir > 1.1, University of Berkley) or CSF Aß levels [5]. Composite scores were calculated for visuospatial, memory, executive function, and language domains [6] from the ANDI2/GO battery. No demographic group effects were observed. All neurocognitive domain scores exhibited a significant group effect (Table 1).

Table 1. Demographics and neurocognitive comparisons of diagnostic groups.

Variable	CN$_{Aβ-}$ (n = 14)	CN$_{Aβ+}$ (n = 12)	EMCI$_{Aβ+}$ (n = 22)	LMCI$_{Aβ+}$ (n = 13)	AD$_{Aβ+}$ (n = 21)
Age (Years) (SD)	74.2 (8.8)	75.9 (7.0)	72.6 (5.2)	73.3 (6.1)	73.5 (7.6)
Sex (% F)	64.2	41.7	50	61.6	42.9
Years of education (SD)	16.7 (2.3)	15.8 (2.6)	15.2 (2.6)	16 (1.8)	15.4 (2.6)
Visuospatial domain score (SD)**	9.7 (0.61)	9.3 (0.9)	9.4 (0.9)	83 (2.3)	7.4 (2.1)
Language domain score (SD)**	49.2 (4.2)	48.8 (4.4)	46.2 (5.8)	43.1 (8.0)	34.8 (9.6)
Memory domain score (SD)**	125.4 (41.1)	142 (34.5)	104.9 (46.6)	81.0 (36.7)	34.2 (21.8)
Executive function domain score (SD)**	99.0 (26.8)	117.6 (27.4)	135.0 (48.6)	166.3 (102.0)	284.6 (101.0)

**Significant group effect (Chi-square or ANOVA as appropriate, α = 0.05)

2.2   fMRI Data Processing

MRI scans used for construction of FC matrices included T1-weighted MPRAGE scans and EPI fMRI scans from the initial visit in ADNI2/GO (www.adni-info.org for protocols). fMRI scans were processed in MATLAB using an FSL based pipeline following processing guidelines by Power et al. [7] and described in detail in Amico et al. [8]. Subjects with over 30% of volumes censored due to motion were discarded to ensure data quality. For purposes of denoising FC matrices [4], processed fMRI time series were split into halves, representing "test" and "retest" sessions.

2.3   Test-Retest Identifiability and Construction and of Individual FC Matrices

For each subject, two FC matrices were created from the "test" and "retest" halves of the fMRI time-series. FC nodes were defined using a 286 region parcellation [9], as detailed in Amico et al. [8]. Functional connectivity matrices were derived by calculating the pairwise Pearson correlation coefficient (r_{ij}) between the mean fMRI time-series of all nodes. "Test" and "retest" FCs were de-noised by using group level PCA to maximize test-retest differential identifiability (I_{diff}) [4]. The "identifiability matrix" I was defined as the matrix of pairwise correlations (square, non-symmetric) between the subjects' FC_{test} and FC_{retest}. The dimension of I is N^2, where N is the number of subjects in the cohort. Self-identifiability, (I_{self}, Eq. 1), was defined to be the average of the main diagonal elements of I, consisting of correlations between FC_{test} and FC_{retest} from the same subjects. I_{others} (Eq. 2), was defined as average of the off-diagonal elements of matrix I, consisting of correlations between FC_{test} and FC_{retest} of different subjects. Differential identifiability (I_{diff}, Eq. 3) was defined as the difference between I_{self} and I_{others}.

$$I_{self} = \frac{1}{N} \sum_{i=j} I_{i,j} \tag{1}$$

$$I_{others} = \frac{1}{N} \sum_{i \neq j} I_{i,j} \tag{2}$$

$$I_{diff} = 100 * \left(I_{self} - I_{others} \right) \tag{3}$$

Group level PCA [10] was applied in the FC domain, on a data matrix ($\mathbf{Y_1}$) containing vectorized FC_{test} and FC_{retest} (upper triangular) from all subjects. PCs throughout this paper will be numbered in order of variance explained. The number of PCs estimated was constrained to 2 * N, the rank of the data matrix $\mathbf{Y_1}$. Following decomposition, PCs were iteratively added in order of variance explained. Denoised FC_{test} and FC_{retest} matrices were reconstructed using the number of PCs (n) that maximized I_{diff} (Eq. 3), while maintaining a minimum I_{others} value of 0.4, such that between-subject FC was neither overly correlated (loss of valid inter-subject variability) nor overly orthogonal (inter-subject variability dominated by noise). This was done because the ADNI2/GO fMRI data was noisier than data on which this method was previously implemented, as evidenced by a much lower original between-subject FC correlation (I_{others} 0.22 ADNI vs. 0.4 Human Connectome Project rs-fMRI [14]). Therefore, not setting a minimum threshold for I_{others} led to the algorithm picking PCs that were "specialized" to specific subjects. The threshold 0.4 was specifically chosen because it reflected average I_{others} values seen in FCs from previous data, on which this method was implemented [4].

Final, de-noised FC matrices were computed as the average of FC_{test} and FC_{retest}. Nodes were assigned to 9 resting state subnetworks (RSN/RSNs), visual (VIS), somato-motor (SM), dorsal attention (DA), ventral attention (VA), limbic (L), fronto-parietal (FP), and default mode network (DMN) [11] with the additional subcortical (SUB) and cerebellar (CER).

2.4   Diagnostic Identifiability

With the goal of early detection in mind, we hypothesized that FC in non-dementia groups would become significantly less identifiable from FC in $AD_{A\beta+}$ with increased diagnostic proximity to $AD_{A\beta+}$. Figure 1 delineates the work flow for finding AD sensitive PCs using a novel diagnostic identifiability metric (D), which quantifies differentiability in connectivity between each non-dementia group (g) and $AD_{A\beta+}$ and is calculated from the correlation matrix ($\mathbf{I}$) of $\mathbf{Y_2}$. D_g was defined as the average correlation within a non-dementia group, corr(g, g), minus the average correlation between that non-dementia group and $AD_{A\beta+}$, corr(g, $AD_{A\beta+}$). D, rather than variance explained, was used to filter components, as it was hypothesized that early disease changes likely do not account for a large portion of between subject variance.

$$D_g = corr(g, g) - corr\left(g, AD_{A\beta+}\right) \tag{4}$$

Group level PCA was again performed on the matrix $\mathbf{Y_2}$. Here, the number of PCs was constrained to $n = 35$ PCs, the rank of the $\mathbf{Y_2}$ matrix. $\mathbf{Y_2}$ was iteratively reconstructed using a subset of the n PCs, selected based on maximizing D_g. Starting with PC$_1$, PC$_{2...n}$ were iteratively added based on their influence on average (D_g). At each iteration, the PC$_{j*}$ which most improved mean (D_g) upon its addition to previously selected PCs, was selected. To avoid results driven by a subset of the population or by differences in sample sizes between groups, the cohort was randomly sampled 30 times, following total cohort PCA, in a group balanced fashion ($n_{sample} = 50$; $n_g = 10$). The number of bootstraps was chosen to allow adequate estimation of the D_g distribution while keeping run-time of the algorithm, reasonable. Bootstrapped distributions of D_g were generated for each number of PCs. The number of PCs (n*) which maximized average (D_g) was found. AD sensitive PCs were defined as those which appeared within the n* most influential PCs with the greatest frequency across samples. Final FC matrices were re-constructed using only AD sensitive PCs.

$$n^* = n, \, at \, argmax_n \left(\frac{1}{g} \sum_g D_g \right) \tag{5}$$

Additionally, D_g curves were estimated and disease sensitive PCs were identified for the 9 RSNs individually, by calculating I_{RSN} using the subset of connections where at least one of the nodes in the connection was part of the RSN.

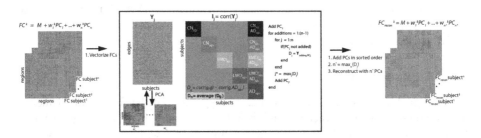

Fig. 1. Diagnostic identifiability workflow.

2.5   Statistical Validation and Association with Neurocognitive Outcomes

Due to the small number of bootstraps, differences between D_g distributions were assessed at n* PCs by checking if the median of one distribution was an outlier relative to a reference distribution using non-parametric confidence intervals defined with the median and interquartile range (IQR). First, D_g distributions from each RSN were compared to those from WB. Next, WB and RSN D_g distributions were compared to a corresponding *null* model. *Null* models for the WB and each RSN were constructed by randomly permuting diagnostic group membership among individuals selected at each bootstrap, such that D_g for the *null* model represented identifiability of a random heterogeneous group from a random heterogeneous reference group. Finally, individual D values (D_i) were calculated for each subject using FC reconstructed with the n* PCs. ANOVA ($\alpha < 0.05$) with follow up pairwise tests, was performed on WB D_i distributions to test for a group effect. Stepwise regressions (F-test, $\alpha = 0.05$), starting with gender, age and education, were be used to test for associations between the n* PC weights and longitudinal changes in neurocognitive outcomes (0, 1, 2 years post imaging).

3   Results

3.1   Test-Retest Identifiability

Figure 2 details the results of denoising FC using differential test-retest differential identifiability. An optimal reconstruction based on the first n = 35 PCs (in decreasing order of explained variance) was chosen (Fig. 2A). I_{self} increased from 0.52 to 0.92

(Fig. 2A–B) while I_{others} increased from 0.20 to 0.40 (Fig. 2A–B). I_{diff} increased from 38% to 57% (Fig. 2A–B).

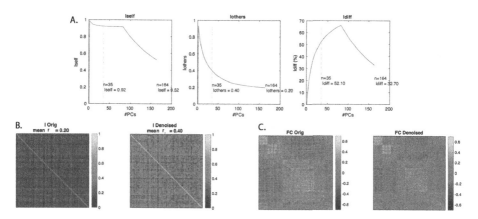

Fig. 2. (A) I_{self}, I_{others}, and I_{diff} across the range of # PCs. (B) I matrices for original and denoised FC matrices. (C) Example original FC matrix versus denoised FC matrix.

3.2 Diagnostic Identifiability

WB average (D_g) peaked at n* = 11 components which explained 58.82% of the variance in the denoised FC data (Fig. 3A, Table 2). At n* PCs, $LMCI_{A\beta+}$ was the only group who that did not exhibit significantly increased D_g from the *null* model. At n* components, D_i distributions exhibited a significant group effect. D_i decreased with diagnostic proximity to $AD_{A\beta+}$ (Fig. 3B). Between-subject correlation in FC increased from 0.41 to 0.71 after reconstruction with n* PCs (Fig. 3B). Of the 9 RSNs, the L network exhibited significantly greater D_{RSN} as compared to WB (Table 2). Like WB, $LMCI_{A\beta+}$ was the only group that did not exhibit significantly greater RSN D_g than the null model, with the exceptions of SM where $EMCI_{A\beta+}$ was additionally not significantly different from the *null* model and L where all non-dementia groups exhibited greater D_g than the *null* model (Table 2). Eight of eleven PCs were identified as disease sensitive in all 9 RSNs and WB (Table 2).

Four PCs exhibited significant associations with various neurocognitive domain scores (Table 3). Visuospatial domain scores were associated with PC 17 at 1 year post imaging and PC 9 at 2 years post imaging. Memory domain scores were associated with PC 32 at 1 year post imaging and PC 7 at 2 years post imaging. Language domain scores were associated with PC 23 at 0 year post imaging and PC 7 at 1 years post imaging. Finally, PC 17 was associated with executive domain scores at 1 and 2 years post imaging.

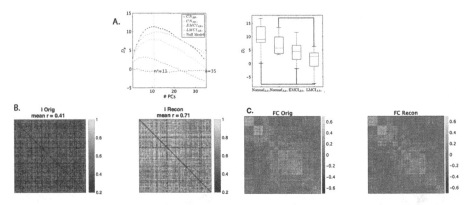

Fig. 3. (A-left) Whole brain D_g across all possible number of PCs. (A-right) Individual D_g values at n* = 11 PCs. Distributions showing significant differences (t-test, $p < 0.05$) are delineated using lines. (B) Original I matrix versus I matrix reconstructed using disease sensitive PCs. (C) Example original FC matrix versus FC matrix reconstructed using disease sensitive PCs.

Table 2. Diagnostic identifiability summary.

RSN	CN$_{A\beta-}$	CN$_{A\beta+}$	EMCI$_{A\beta+}$	LMCI$_{A\beta+}$	Mean	n	Var (%)
WB	11.35**	9.75**	7.85**	2.60	7.89	11	58.82
VIS	13.21**	10.33**	8.11**	2.75	8.60	10	57.23
SM	9.82**	12.96**	7.30	4.43	8.62	10	57.26
DA	12.16**	11.24**	8.09**	3.37	8.71	13	62.19
VA	10.74**	11.65**	7.96**	2.33	8.17	10	57.26
L	**17.18****	**13.76****	**11.97****	**6.28****	**12.30**	8	54.50
FP	12.07**	10.25**	9.34**	2.70	8.59	11	58.82
DMN	12.09**	10.20**	8.39**	2.84	8.38	11	58.82
SUB	14.17**	11.66**	9.93**	4.33	10.02	9	55.78
CER	13.29**	12.85**	9.72**	5.67	10.38	10	57.26

**Median outside CI null model, Median outside CI WB mean (D_g)

Table 3. Associations of n* PC weights with neurocognitive composite domain scores. Stepwise regressions (F-test, $\alpha < 0.05$) were used to assess the relationship of neurocognitive composite domain scores with PC weights, with age, gender, and education starting in the base model; p values are reported for the whole model, adjusted-R^2 is reported for the model.

Time points	Visuospatial			Memory			Language			Executive		
	PC	p	R^2	PC	p	R^2	PC	p	R^2	PC	p	R^2
0	–	–	–	–	–	–	23	0.040	0.19	–	–	–
1	17	0.001	0.53	32	0.032	0.31	7	0.025	0.31	17	0.004	0.48
2	9	0.020	0.46	7	0.044	0.20	–	–	–	17	0.013	0.36

4 Limitations, Future Work, and Conclusions

We present here a two stage PCA based framework to improve the detection of AD signatures in whole-brain functional connectivity. We first use recently proposed test-retest differential identifiability to denoise subject-level functional connectomes and consequently reduce dimensionality of functional connectomes. We subsequently introduce and validate the concept of PCA based differential diagnostic identifiability to increase AD signal to background in functional connectivity. The result of a significant diagnostic group effect in diagnostic differential identifiability shows that FC contains AD signature, even at early stages of disease. The finding of increased diagnostic identifiability in Limbic regions, known to be associated with memory processes and known to be affected in AD, further validates this finding. Finally, we show that PC weights from AD sensitive principal components are correlated to longitudinal neurocognitive outcomes. In addition to the work presented here, we plan to delve further into the meaning of the PCs themselves. AD sensitive PCs did not appear to be specific to individual RSNs, as the same PCs were consistently AD sensitive across RSNs. Furthermore, several PCs were associated with multiple neurocognitive domains. Therefore, AD sensitive PCs may characterize global brain changes related to AD. However, spatial representation of PCs and relationship of PCs with network properties need to be explored to further assess this. Finally, to further validate these promising results, this methodology needs to be applied to a larger cohort. With ADNI3 data becoming available (~ 300 subjects already scanned), on which all subjects underwent resting state fMRI, we will be able to further validate findings and further improve identification and characterization of AD sensitive PCs based on whole brain functional connectomes. This dual decomposition/reconstruction framework makes forward progress in exploiting the clinical potential of functional connectivity based biomarkers.

References

1. Sperling, R.A., Karlawish, J., Johnson, K.A.: Preclinical Alzheimer disease - the challenges ahead. Nat. Rev. Neurol. **9**(1), 54–58 (2013)
2. Brier, M.R., Thomas, J.B., Ances, B.M.: Network dysfunction in Alzheimer's disease: refining the disconnection hypothesis. Brain Connect. **4**(5), 299–311 (2014)
3. Braun, U., et al.: Test - retest reliability of resting-state connectivity network characteristics using fMRI and graph theoretical measures. Neuroimage **59**(2), 1404–1412 (2012)
4. Amico, E., Goni, J.: The quest for identifiability in human functional connectomes. Sci. Rep. **8**(1), 8254 (2018)
5. Shaw, L.M., et al.: Cerebrospinal fluid biomarker signature in Alzheimer's disease neuroimaging initiative subjects. Ann. Neurol. **65**(4), 403–413 (2009)
6. Wilhalme, H., et al.: A comparison of theoretical and statistically derived indices for predicting cognitive decline. Alzheimers Dement. (Amst) **6**, 171–181 (2017)
7. Power, J.D., et al.:Spurious but systematic correlations in functional connectivity MRI networksarise from subject motion. Neuroimage, **59**(3): 2142–54, (2012)
8. Amico, E., et al.: Mapping the functional connectome traits of levels of consciousness. Neuroimage **148**, 201–211 (2017)

9. Shen, X., et al.: Groupwise whole-brain parcellation from resting-state fMRI data for network node identification. Neuroimage **82**, 403–415 (2013)
10. Hotelling, H.: Analysis of complex variables into principal components. J. Educ. Psychol. **24**, 417–441 (1933)
11. Yeo, B.T.T., et al.: Estimates of segregation and overlap of functional connectivity networks in the human cerebral cortex. NeuroImage **88**, 212–227 (2014)

Predicting Conversion of Mild Cognitive Impairments to Alzheimer's Disease and Exploring Impact of Neuroimaging

Yaroslav Shmulev[1,2(✉)], Mikhail Belyaev[1,2],
and the Alzheimer's Disease Neuroimaging Initiative

[1] Kharkevich Institute for Information Transmission Problems, Moscow, Russia
yaroslav.shmulev@skolkovotech.ru
[2] Skolkovo Institute of Science and Technology, Moscow, Russia

Abstract. Nowadays, a lot of scientific efforts are concentrated on the diagnosis of Alzheimers Disease (AD) applying deep learning methods to neuroimaging data. Even for 2017, there were published more than hundred papers dedicated to AD diagnosis, whereas only a few works considered a problem of mild cognitive impairments (MCI) conversion to AD. However, the conversion prediction is an important problem since approximately 15% of patients with MCI converges to AD every year. In the current work, we are focusing on the conversion prediction using brain Magnetic Resonance Imaging and clinical data, such as demographics, cognitive assessments, genetic, and biochemical markers. First of all, we applied state-of-the-art deep learning algorithms on the neuroimaging data and compared these results with two machine learning algorithms that we fit on the clinical data. As a result, the models trained on the clinical data outperform the deep learning algorithms applied to the MR images. To explore the impact of neuroimaging further, we trained a deep feed-forward embedding using similarity learning with Histogram loss on all available MRIs and obtained 64-dimensional vector representation of neuroimaging data. The use of learned representation from the deep embedding allowed to increase the quality of prediction based on the neuroimaging. Finally, the current results on this dataset show that the neuroimaging does have an effect on conversion prediction, however cannot noticeably increase the quality of the prediction. The best results of predicting MCI-to-AD conversion are provided by XGBoost algorithm trained on the clinical and embedding data. The resulting accuracy is ACC $= 0.76 \pm 0.01$ and the area under the ROC curve – ROC AUC $= 0.86 \pm 0.01$.

Data used in preparation of this article were obtained from the Alzheimer's Disease Neuroimaging Initiative (ADNI) database (adni.loni.usc.edu). As such, the investigators within the ADNI contributed to the design and implementation of ADNI and/or provided data but did not participate in analysis or writing of this report. A complete listing of ADNI investigators can be found at: http://adni.loni.usc.edu/wp-content/uploads/how_to_apply/ADNI_Acknowledgement_List.pdf.

© Springer Nature Switzerland AG 2018
D. Stoyanov et al. (Eds.): GRAIL 2018/Beyond MIC 2018, LNCS 11044, pp. 83–91, 2018.
https://doi.org/10.1007/978-3-030-00689-1_9

Keywords: Image classification · Similarity learning
Disease progression · CNN · MRI

1   Introduction

Alzheimer's Disease is irreversible progressive brain disorder mostly occurring in the middle or late life. At the same time, there is a transitional phase between the normal aging and dementia symptoms called mild cognitive impairment (MCI). People with MCI are at increased risk of AD development – approximately 15% of them converge to dementia every year. That's why, the early diagnosis of Alzheimer's Disease would allow patients to take preventive measures to temporarily slow the disease progression [10].

Neuroimaging is a variety of methods and technologies that reveal the structure and functions of brain. It includes Computer Tomography (CT), structural and functional Magnetic Resonance Imaging (sMRI and fMRI respectively) and etc. With a growth of deep learning applications in data analysis, neuroimaging is extensively used in many medical tasks such as image segmentation [1], diagnosis classification [11] and prediction of disease progression [5].

In the recent years, there were published a vast number of papers dedicated to classification of healthy controls from AD using deep learning approach applied to neuroimaging. However, only a few works considered predicting conversion of MCI to AD [5,6,8], which is a more complicated and clinically relevant problem. To classify stable and converged MCI the authors of [6] used different clinical biomarkers and complex feature maps extracted from neuroimaging. This method inherit the main drawbacks from manual feature extraction procedure. Cheng et al. in [5] consider the joint multi-domain learning for early diagnosis of AD to boost the learning performance. In this work, we are focusing on the conversion prediction using clinical and neuroimaging data. In addition, we want to explore the individual impact of different data types on the prediction performance for different prediction intervals. Finally, we obtain low-dimensional representation of high-dimensional MR brain images from a deep feed-forward embedding that is trained on the whole ADNI cohort.

2   Data

In this work, we use data obtained from the Alzheimer's Disease Neuroimaging Initiative (ADNI) database [2]. We choose patients that are diagnosed as normal controls (NC), mild cognitive impairments (MCI), and Alzheimer's Disease (AD). For each patient, we take visits for which both MR images and clinical data are available. The total number of available data samples is 8809.

Clinical Data. ADNI provides clinical information about each subject including recruitment, demographics, physical examinations, and cognitive assessment data. We add genetic and biospecimen data (cerebrospinal fluid concentration, blood, and urine) to the clinical dataset. The full list of attributes is available on the official ADNI website [2].

Neuroimaging Data. For the analysis, we take structural T1-weighted Magnetic Resonance Imaging (MRI), since they are available for all patients and for the most their visits. We fetch preprocessed images from ADNI database with the following preprocessing pipeline descriptions: **"MPR; GradWarp; B1 Correction; N3; Scaled"** and **"MT1; GradWarp; N3m"**. These images have different shapes and orientations and contain skull and other organs that might spoil a predictive performance. Thus, we apply the following preprocessing pipeline for the collected neuroimaging dataset. For the Brain extraction [3] and N4 bias correction [13] steps, we run ANTs Cortical Thickness Pipeline [14] for all available MR images. Then, we apply an affine transform, so that all brain images have the same orientation - **RAS** (Left - Right, Posterior - Anterior, Superior - Inferior). After the brain extraction step, the MRIs contain a lot of black voxels around the brain. We crop all images to the maximal extracted brain size, which is computed beforehand. Ultimately, all MR images in the dataset have a size of $(150, 208, 173)$. To increase a batch size that can be fitted to the Graphics Processing Unit (GPU), we downsample the dataset with the factor of 2 for each dimension, so that the resulting shape is $(75, 104, 87)$.

Conversion Dataset. To predict the MCI-to-AD conversion, we need to remove patients that are normal controls (NC) or have Alzheimer's Disease from the screening visit.

For the stable MCI, we consider participants that have not converged to AD for the known time-period. We also drop several last visits that are inside the prediction horizon, since the future for that patients is not known and they may converge to AD in the next visits.

For the converged MCI, we select participants that were diagnosed as MCI in earlier sessions and as AD later. We take visits that are within five year prediction interval. The total number of stable and converged patients are 532 and 327 correspondingly. The number of samples for two classes are 1764 and 1016.

3   Method

3.1   Clinical Data

For the classification based on clinical data, we use two machine learning algorithms: Logistic Regression and XGBoost. The first one is a linear method which is widely used in many practical applications because of its good interpretability and relative simplicity. The second method is an efficient implementation of gradient boosting on decision trees, which is a powerful machine learning algorithm that can catch nonlinear patterns in data [4].

3.2   Neuroimaging Data

Convolutional Neural Networks (CNN) have recently made a great breakthrough in the image classification and recognition tasks. Deep CNNs automatically extract and combine from low- to high-level features from images and estimate target values in the end-to-end fashion. In this work, we use two deep architectures: VGG [12] and ResNet [7], that showed state-of-the-art performance in ImageNet classification challenge in 2014 and 2015 correspondingly. We generalize these architectures to the three-dimensional input size of MR images in the same way as was proposed in [11].

VoxCNN. The VGG-like network consists of ten 3D convolutional blocks, each of which consists of three 3D-convolutional layers with $3 \times 3 \times 3$ filter sizes, batch normalization and ReLU nonlinearity. Then, we use max pooling layer with $2 \times 2 \times 2$ kernel size to reduce the size of data propagated through the network. At the end of the net, there are three fully-connected layers with batch normalization and dropout layers in-between. For the experiments, we used the probability $p = 0.7$ of a neuron to be turned off. After the last fully-connected layer, there is softmax activation function to compute probabilities for each class.

ResNet3D. For the ResNet-like architecture, we use 6 residual blocks, each of which represents a sum of identity mapping and a stack of two convolutions with $3 \times 3x3$ filter size and 64 or 128 filters, batch normalization and ReLU. The convolutional layers from the standard ResNet are replaced with 3D ConvBlocks in the same way as we did for VoxCNN. We reduce the spacial size using three convolutions with strides $2 \times 2 \times 2$ before the residual blocks and one maximum pooling layer with $5 \times 5 \times 5$ kernel size before the fully-connected layer. Dropout with $p = 0.7$ and batch normalization are also used after the first fully-connected layer. At the end of the network, there is a second fully-connected layer with softmax activation to produce output probabilities.

4 Experiments

4.1 Setup

For the experiments with conversion prediction based on the neuroimaging data, we minimize a weighted binary cross-entropy loss function with balanced class weights. We use Nesterov momentum optimizer with initial learning rate 10^{-3} and scheduling learning rate policy: we decrease the learning rate ten-fold after 30 and 50 epochs. The batch sizes for ResNet3D and VoxCNN are 128 and 512 correspondingly. These numbers are chosen so that the full batch can be fitted to the GPU. The total number of epochs is 70.

4.2 Validation

To assess the classification performance more accurately, we run 5-fold group cross-validation with five different folds. As a group label, we use participant's ID to prevent the cases when different scans of one patient are simultaneously in train and test sets.

For neuroimaging data, on each step of cross-validation procedure, we train a separate neural network on a train set, use a validation set for early stopping and changing learning rate, and test the network model on a hold-out subset.

For hyperparameter tuning of Logistic Regression and XGBoost methods, another nested group cross-validation procedure is used.

We report the following metrics: accuracy, the area under the receiver operating characteristic curve (ROC AUC), sensitivity, specificity, and average precision.

4.3 Conversion Prediction

There are several ways how the disease progression problem can be formulated. In this work, we use binary classification setting to predict the fact of conversion within a five year interval: **class 0** - stable MCI, **class 1** - converged MCI. In other words, given a participant's visit, we would like to answer the question, whether an individual will converge to AD within the considered interval or not.

4.4 Embedding Learning

For conversion prediction, we used only 25% of all available MR images. To make use of all available data, we learn a deep feed-forward embedding on the whole neuroimaging dataset and, then, use it as a fixed feature extractor. We exploit the extracted features for conversion prediction task, as shown in Fig. 1.

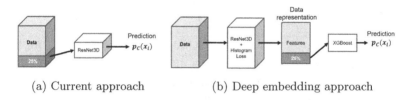

(a) Current approach   (b) Deep embedding approach

Fig. 1. Two approaches for conversion prediction task. (a) In the current approach, only 25% of all available MR images are used for the conversion prediction. (b) The embedding is trained on the whole MRI dataset and, then, used for feature extraction. We use the extracted features for the conversion prediction task.

Deep embedding is an approach, when complex high-dimensional input data are mapped into a smaller size semantic subspace preserving the most relevant information about the data. Generally, the mapping is learned from a large amount of supervised data. During the training process, semantically related samples are getting closer than semantically unrelated ones in the semantic subspace. To learn deep feed-forward embedding we use ResNet3D architecture up to the last fully connected layer. We add a fully connected layer with 64 output units and L2-normalization layer to the network. We use Histogram loss proposed in [15] as a training criteria, which is parameter-free batch loss function that firstly estimates two distributions of distances between matching and non-matching pairs and, secondly, computes the overlap between these two distributions. Once the deep embedding is trained, we use it to extract the embedded features: all images from the conversion dataset are propagated through the network and 64-dimensional vector representations are obtained.

5   Results

The results of the conversion prediction are shown in Table 1. For the considered interval, the quality of prediction based on clinical data is significantly higher than one on the neuroimaging. On the neuroimaging, two network architectures provide comparative results for all experiments, although ResNet3D slightly outperforms VoxCNN. For the clinical data, the performance of XGBoost is slightly better than one of Logistic Regression model.

The results also show that the use of learned deep embedding helps increase the quality of prediction based on MR images, although it is still worse than one on the clinical data.

To investigate whether the neuroimaging data can add some new relevant information to the clinical data and, thereby, improve the prediction, we include extracted features from the embedding to the clinical ones. As can be seen from the results, the quality of prediction using clinical and embedding data is slightly higher than for clinical data, although still it is the same within the standard deviation.

Table 1. Conversion prediction results

Data/Method	ACC	ROC AUC	AV PREC	SENS	SPEC
Clinical data/Log Reg	.76 ± .01	.85 ± .01	.73 ± .05	.80 ± .03	.74 ± .02
Clinical data/XGBoost	.76 ± .01	.85 ± .01	.73 ± .03	.76 ± .02	.77 ± .01
Neuroimaging/VoxCNN	.61 ± .02	.70 ± .03	.52 ± .05	.70 ± .04	.56 ± .02
Neuroimaging/ResNet3D	.62 ± .01	.70 ± .02	.53 ± .02	.75 ± .03	.54 ± .01
Embedding/Log Reg	.69 ± .01	.71 ± .01	.54 ± .03	.60 ± .01	.75 ± .03
Embedding/XGBoost	.67 ± .02	.73 ± .01	.57 ± .02	.70 ± .02	.65 ± .05
Clinic. + Embed./Log Reg	.76 ± .02	.86 ± .02	.73 ± .03	.84 ± .02	.72 ± .03
Clinic. + Embed./XGBoost	.76 ± .01	.86 ± .01	.73 ± .02	.88 ± .03	.70 ± .03

Figure 2 shows the resulting representation from the learned deep embedding on a hold-out set. We applied T-SNE algorithm proposed in [9] to map our 64-dimensional feature vectors into 2-dimensional ones. In Fig. 2a, there are three clusters, each of which corresponds to one of the diagnoses: NC, MCI, or AD. From Fig. 2b and d can be seen that the separation between NC and AD is better than separation between MCI and AD.

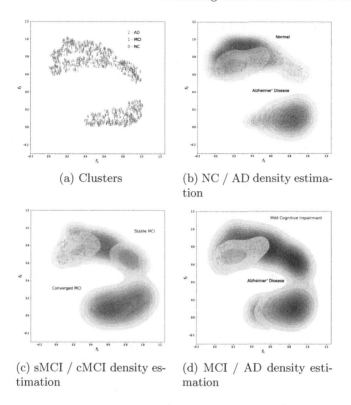

(a) Clusters

(b) NC / AD density estimation

(c) sMCI / cMCI density estimation

(d) MCI / AD density estimation

Fig. 2. Embedding visualization: (a) clusters in the embedded space, (b) Kernel Density Estimation (KDE) of NC and AD distributions, (c) KDE of stable and converged MCI distributions, (d) KDE of MCI and AD distributions.

The next observation from Fig. 2a is that MCI cluster is spread between normal controls (NC) and Alzheimer's Disease (AD). Since we know which MCI patient will converge to AD (cMCI) and which will not (sMCI), we plot the densities of stable and converged MCI. Figure 2c shows that these two groups of MCIs are quite good separated in the embedded space. The main mass of converged MCIs is closer to the AD cluster, whereas the stable MCIs are closer to the normal controls.

6   Conclusion

In this work, a problem of conversion prediction from mild cognitive impairment (MCI) to Alzheimer's Disease (AD) was considered. We collected, preprocessed and analyzed the clinical and neuroimaging data. We applied the state-of-the-art methods for image classification on the neuroimaging data and compared the quality of classification with the several machine learning methods trained on the clinical data. The results of the experiments showed that the clinical data allow to obtain a better prediction quality than the neuroimaging and these models can be used for conversion prediction task.

We enhanced the performance on the neuroimaging data by training a deep feed-forward embedding. The embedding increased the quality of forecast, however, it is still worse than the clinical data yield. We further investigate the question whether the neuroimaging is able to add some new information for conversion prediction or not. According to the results on the current dataset, neuroimaging does have an effect on the conversion prediction, however it cannot noticeably increase the quality of the prediction when clinical data are used.

Acknowledgements. The obtained results has been obtained under support of the Russian Science Foundation grant 17-11-0139.

References

1. Akkus, Z., Galimzianova, A., Hoogi, A., Rubin, D.L., Erickson, B.J.: Deep learning for brain MRI segmentation: state of the art and future directions. J. Digit. Imaging **30**(4), 449–459 (2017)
2. Alzheimer's Disease Neuroimaging Initiative (2003). http://adni.loni.usc.edu/. Accessed 22 May 2018
3. Avants, B., et al.: Evaluation of an open-access, automated brain extraction method on multi-site multi-disorder data. In: 16th Annual Meeting for the Organization of Human Brain Mapping (2010)
4. Chen, T., Guestrin, C.: XGBoost: a scalable tree boosting system. In: Proceedings of the 22nd ACM SIGKDD International Conference on Knowledge Discovery and Data Mining, pp. 785–794. ACM (2016)
5. Cheng, B., Liu, M., Shen, D., Li, Z., Zhang, D., Alzheimer's Disease Neuroimaging Initiative: Multi-domain transfer learning for early diagnosis of Alzheimer's disease. Neuroinformatics **15**(2), 115–132 (2017)
6. Davatzikos, C., Bhatt, P., Shaw, L.M., Batmanghelich, K.N., Trojanowski, J.Q.: Prediction of MCI to AD conversion, via MRI, CSF biomarkers, and pattern classification. Neurobiol. Aging **32**(12), 2322–e19 (2011)
7. He, K., et al.: Deep residual learning for image recognition. In: Computer Vision and Pattern Recognition, December 2015
8. Hu, K., Wang, Y., Chen, K., Hou, L., Zhang, X.: Multi-scale features extraction from baseline structure MRI for MCI patient classification and AD early diagnosis. Neurocomputing **175**, 132–145 (2016)
9. van der Maaten, L., Hinton, G.: Visualizing data using t-SNE. J. Mach. Learn. Res. **9**(Nov), 2579–2605 (2008)
10. Roberson, E.D., Mucke, L.: 100 years and counting: prospects for defeating Alzheimer's disease. Science **314**(5800), 781–784 (2006)
11. Sarraf, S., Tofighi, G.: Classification of Alzheimer's disease structural MRI data by deep learning convolutional neural networks. arXiv preprint arXiv:1607.06583 (2016)
12. Simonyan, K., Zisserman, A.: Very deep convolutional networks for large-scale image recognition. arXiv preprint arXiv:1409.1556 (2014)
13. Tustison, N.J., et al.: N4ITK: improved N3 bias correction. IEEE Trans. Med. Imaging **29**, 1310–1320 (2010)

14. Tustison, N.J., et al.: The ANTs longitudinal cortical thickness pipeline. In: Proceedings of SPIE (2013)
15. Ustinova, E., Lempitsky, V.: Learning deep embeddings with histogram loss. In: Advances in Neural Information Processing Systems, pp. 4170–4178 (2016)

Cross-diagnostic Prediction of Dimensional Psychiatric Phenotypes in Anorexia Nervosa and Body Dysmorphic Disorder Using Multimodal Neuroimaging and Psychometric Data

Jamie D. Feusner[1,2]([⊠]), Wesley T. Kerr[1,2,3], Teena D. Moody[1,2], Aifeng F. Zhang[4], Mark S. Cohen[1,2], Alex D. Leow[4,5], Michael A. Strober[1,2], and Don A. Vaughn[1,6]

[1] Semel Institute for Neuroscience and Human Behavior, University of California, Los Angeles, CA, USA
JFeusner@mednet.ucla.edu
[2] David Geffen School of Medicine, University of California, Los Angeles, CA, USA
[3] Department of Internal Medicine, Eisenhower Health, Rancho Mirage, CA, USA
[4] Department of Psychiatry, University of Illinois, Chicago, IL, USA
[5] Department of Bioengineering, University of Illinois, Chicago, IL, USA
[6] Department of Psychology, Santa Clara University, Santa Clara, CA, USA

Abstract. Anorexia nervosa (AN) and body dysmorphic disorder (BDD) share several phenomenological features including distorted perception of appearance, obsessions/compulsions, and limited insight. They also show partially over-lapping patterns of brain activation, white matter connectivity, and electro-physiological responses. These markers have also shown associations with symptom severity within each disorder. We aimed to determine: (a) if, cross-diagnostically, neural activity and connectivity predict dimensional clinical phenotypes, and (b) the relative contribution of multimodal markers to these predictions beyond demographics and psychometrics, in a multivariate context. We used functional magnetic resonance imaging (fMRI) data from a visual task, graph theory metrics of white matter connectivity from diffusor tensor imaging, anxiety and depression psychometric scores, and demographics to predict dimensional phenotypes of insight and obsession/compulsions across a sample of unmedicated adults with BDD (n = 29) and weight-restored AN (n = 24). The multivariate model that included fMRI and white matter connectivity data performed significantly better in predicting both insight and obsessions/compulsions than a model only including demographics and psychometrics. These results demonstrate the utility of neurobiologically-based markers to predict important clinical phenotypes. The findings also contribute to understanding potential cross-diagnostic substrates for these phenotypes in these related but nosologically discrete disorders.

Keywords: fMRI · DTI · Insight · Obsessions · Compulsions
Multivariate

© Springer Nature Switzerland AG 2018
D. Stoyanov et al. (Eds.): GRAIL 2018/Beyond MIC 2018, LNCS 11044, pp. 92–99, 2018.
https://doi.org/10.1007/978-3-030-00689-1_10

1 Introduction

Anorexia nervosa (AN) and body dysmorphic disorder (BDD) are psychiatric disorders with a high risk of morbidity and mortality [1]. Core symptoms of AN include reduced caloric intake, low body weight, fear of becoming fat, and disturbed experience of one's body or weight; in BDD these include preoccupation with perceived defects in appearance and repetitive behaviors to check, fix, change, or hide aspects of their appearance [1].

AN and BDD are categorized as an eating disorder and as an obsessive-compulsive related disorder, respectively, yet they share phenomenological features such as distorted perception of appearance, poor insight [2, 3], and obsessive and compulsive symptoms [4]. Of those with AN, 25–39% are diagnosed with lifetime BDD; 32% of those with BDD will have a lifetime eating disorder [5, 6]. Additionally, 30% of those with BDD have weight-related appearance concerns (e.g. their cheeks or thighs being too fat) [7]. Similarities have raised the question of whether one disorder should be considered a subtype of the other, or that they share pathological features [8].

The few studies that have directly compared the neurobiology of AN and BDD demonstrate overlapping and distinct patterns of neural activity and connectivity [9, 10]. Several of these studies have also examined associations between neural markers and clinical symptoms. N170 ERP amplitude correlated with insight in BDD but not in AN [10]. Insight was correlated with a graph theory network measure, normalized path length (NPL), in white matter in AN but not in BDD [3, 10]. Studies in BDD have shown associations between obsessions and compulsions and activation in prefrontal, striatal, and visual regions [11]; with connectivity in the orbitofrontal cortex [12]; and with whole-brain white matter connectivity (global efficiency) [13]. Insight in BDD is associated with regional white matter diffusion in tracts relevant to visual processing [14].

Whether common phenotypic symptom profiles are associated with underlying brain activation patterns and white matter structural properties remains unexplored. This is relevant, as psychiatric disorders once assumed to be causally independent have been found to have common genetic variant risks [15]. Such relationships could inform underlying shared or unique neurobiological and brain-behavior relationships contributing to dimensional phenotypes. Accordingly, a goal of this study was to determine if, cross-diagnostically, neural activity and connectivity patterns predict dimensional phenotypes. Potential clinical value of this would be at the cost of obtaining neuroimaging markers, which is not part of standard clinical practice; thus, a second goal was to determine the relative predictive contribution of imaging markers beyond demographic and psychometric data. We hypothesized that neural activity and connectivity patterns would significantly predict insight and obsession/compulsion phenotypes across AN and BDD, and would provide additional significant predictive value beyond demographics and psychometrics, in a multivariate context.

2 Methods

2.1 Participants

Fifty-three individuals participated, between the ages of 14 and 38. Twenty-nine met Diagnostic and Statistical Manual (DSM-IV) [16] criteria for BDD, and 24 for AN, aside from being weight-restored (BMI $\geq$ 18.5) to avoid confounds of starvation state (Table 1).

Table 1. Demographics. Errors are standard deviation.

Variable	AN	BDD	P
Number of participants	24	29	N/A
Age (years)	21 ± 5	23 ± 5	0.17
Sex: female	23/24 (96%)	26/29 (90%)	0.62
Illness duration (log months)	3.7 ± 1.2	4.6 ± 0.7	0.01
BMI	20 ± 2	22 ± 3	0.02
Lowest lifetime BMI	16 ± 2	N/A	N/A

Participants were free from psychoactive medications for at least 8 weeks. For detailed inclusion and exclusion criteria please see our previous publications [3, 9, 10].

2.2 Psychometrics

All received clinician-rated scales: the Brown Assessment of Beliefs Scale (BABS) [17] (higher scores indicate worse insight), the Hamilton Anxiety Rating Scale (HAMA) [18], and the Montgomery-Asberg Depression Scale (MADRS) [19]. To measure obsession and compulsions, the BDD group received the BDD version of the Yale-Brown Obsessive Compulsive Scale (BDD-YBOCS) [20] and the AN group received a version of the Yale-Brown-Cornell Eating Disorder Scale (YBC-EDS) [21] modified to match the BDD-YBOCS on total numbers of items and a single avoidance and a single insight item. To generate a single regression for predicting obsessions/compulsions, we aggregated data from both groups into one outcome variable, the "YBC/BDD-YBOCS." HAMA. and MADRS were correlated ($r = 0.78$) so we collapsed them into one metric, "HAMADRS," by using the first principal component, which explained 82% of the variance.

2.3 Overview and Rationale of Variable Selection

A goal was to create a prediction model to understand multivariate relationships between insight and obsessions/compulsions across AN and BDD, with functional and structural brain measures, psychometrics, and demographics. We used structural (DTI) and functional (fMRI) data, anxiety and depression (HAMADRS), insight

(BABS), and obsessions/compulsions (BDD-YBOCS and modified YBC-EDS) ratings. From DTI, we used NPL to provide a summarized metric of global white matter network connectivity. We included fMRI data from a task of viewing images of bodies, and faces (visual stimuli that are relevant to participants' appearance concerns) and houses (which are unrelated to appearance concerns).

2.4 fMRI Data

We collected fMRI data on a 3T scanner as participants matched high, normal, and low spatial-frequency images of others' bodies, faces, and houses, as previously described [9, 22]. To derive a signal metric per network, we extracted network coherence values from three networks of interest: primary visual (PV), higher order visual (HV), and salience networks [22]. We collected 64 gradient direction diffusion-weighted images, with $b = 1000$ s/mm^2 and one minimally diffusion-weighted scan. Graph theory metrics were calculated from deterministic tractography-derived connectivity matrices using Freesurfer (Martinos Center for Biomedical Imaging, USA) parcellation of T1 images, as previously described [3]. Shortest path length between each pair of nodes was averaged over all nodes to obtain the characteristic path length (CPL). The normalized path length (NPL) is the ratio of observed CPL to the CPL of an identically sized but randomly connected network [23].

2.5 Missing Data Imputation with Multiple Imputation

We addressed missing data using multiple imputation [24, 25]. We had data for 100% of participants for HAMA/MADRS, 85% of the BABS, 100% of DTI, and 68% of fMRI. We conservatively assumed that data were missing completely at random, namely, unrelated to diagnosis and severity of illness. We used a transformed-linear multivariate model to estimate the covariance of variables, with illness duration modeled as log-linear. We chose this imputation strategy as there was insufficient evidence to suggest that non-linear trends existed, and insufficient data to reliably estimate nonlinear terms within each model. Missing values were imputed 20 independent times based on posterior probabilities of the estimate of the missing data using this multivariate transformed-linear model. The initial values were cold-deck imputed and, to improve exploration of the whole parameter space and reduce tendencies to fall into local minima due to the relatively small dataset, the estimated covariance was multiplied by an exponentially decaying dispersion term with magnitude of 1% after 100 iterations. Each imputation consisted of 400 iterations, although most imputed datasets converged within 200 iterations.

2.6   Statistical Modeling

Linear associations of demographic (age, sex, BMI), clinical variables (AN or BDD diagnosis, the log of illness duration), psychometric scores (HAMADRS, and BABS for the YBC/BDD-YBOCS model), and MRI features (NPL; salience, HV, PV coherence values) with the cross-diagnostic clinical phenotypes of BABS and YBC/BDD-YBOCS (separately) were evaluated using multivariate linear regression. Log-likelihood tests evaluated if including MRI-based features significantly improved the model as compared to only demographic and clinical variables; or demographic, clinical variables and psychometric scores. Primary predicted outcomes were BABS and YBC/BDD-YBOCS; imputation and regression modeling of each were performed separately.

3   Results

3.1   BABS

Model predictions using just demographic variables were significantly different from a constant model (deviance difference 280.7, df = 4, p = 10^{-59}). MRI-based features significantly improved the model as compared to just demographic and clinical variables (deviance difference 89.1, df = 5, p = 10^{-17}); and a model including demographic, clinical variables and HAMADRS (deviance difference 83.5, df = 4, p = 10^{-17}, Fig. 1A). The only factor that trended towards individually significant association was group; the BDD group having a 3.8 higher score than AN (SE 2.2, p = 0.08, Fig. 1B).

3.2   YBC/BDD-YBOCS

Model predictions using just demographic variables significantly differed from a constant model (deviance difference 798.7, df = 4, p = 10^{-171}). MRI-based features significantly improved the model compared to including just demographic and clinical variables (deviance difference 550.5, df = 6, p = 10^{-115}); and a model including demographic, clinical variables, and psychometric scores (deviance difference 233.4, df = 4, p = 10^{-49}, Fig. 1C). The only factors that had significant individual associations were group, with the BDD group having a 8.2 higher score (SE 2.7, p = 0.003, Fig. 1D) and HAMADRS with a unit effect of 6.4 (SE 2.5, p = 0.01).

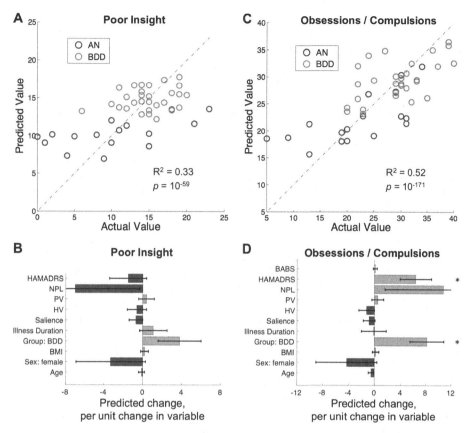

Fig. 1. Model performance and features. A. Performance of the BABS model in predicting participants' observed values. B. Feature weights and errors from the BABS model. C. Performance of the YBC/BDD-YBOCS model in predicting participants' observed values. D. Feature weights and errors from the YBC/BDD-YBOCS model. An asterisk signifies $p < 0.05$.

4   Discussion and Conclusions

A multivariate model that included brain structure and function, psychometrics, and demographics demonstrated significant predictions for both insight and obsessions/compulsions. Moreover, neuroimaging-derived data significantly improved the predictive ability of the model beyond the psychometric and demographic data.

Results suggest that brain structure and function, anxiety and depression, and demographic variables contribute to poor insight across AN and BDD. NPL appeared to contribute more, individually, to predicting insight and obsessions/compulsions than did activation in visual and salience networks. However, inherent to this multivariate analysis is the possibility of complex relationships between variables that do not lend themselves to being disentangled and interpreted in terms of contributions of individual parts.

Models with neurobiological variables were better predictors over the reduced models. This provides early promise that neuroimaging markers might provide clinical utility for predicting dimensional severity of phenotypes across disorders, longitudinally. This requires verification in larger, and longitudinal, studies and those that target specific patient cohorts such as those in early stages of illness or at-risk, to prove pragmatic utility. This is important due to additional costs with neuroimaging.

The sample size limited our ability to include other potentially informative neurobiological and demographic variables. Other statistical modeling approaches—such as training on broader sets of whole-brain activation and connectivity features in a more data-driven manner—could also be applied with larger samples and may improve predictive performance. We had missing data, although we mitigated this by using multiple imputation. We modified the YBC to match the BDD-YBOCS, although the validity and reliability of this modified version has not been tested.

Results shed light on possible shared neurobiological contributors to symptoms in AN and BDD, including white matter network organization indexing long-distance efficiency of brain connections and connectivity within higher- and lower-order visual and salience networks. A tentative model is one in which the combination of specific patterns of visual processing of symptom-related stimuli, combined with specific patterns of white matter network "scaffolding" for how this information integrates across the brain, in the context of anxiety/depression and specific demographics, contributes to worse insight and obsession/compulsion symptoms. A strength of the current analysis is that, as opposed to previous univariate analyses, the functional and structural neurobiological contributors are integrated in a more realistically complex context of variable symptom severity and clinical variables. Our dimensional approach also circumvents limitations of categorical diagnostic categories [26]. This study provides early proof-of-concept for multimodal neurobiological, psychometric, and demographic variables to understand contributors to cross-diagnostic phenotypes and potentially to predict dimensional symptom profiles.

Funding and Disclosure. This work was supported by NIMH grants (R01MH093535 and R01MH105662) to JDF, a Postdoctoral Fellowship to DAV from the UCLA Collaboratory directed by Matteo Pellegrini. The authors declare no conflict of interest in this publication.

References

1. American Psychiatric Association: Diagnostic and Statistical Manual of Mental Disorders (DSM-5®). American Psychiatric Publishing (2013)
2. Hartmann, A.S., Thomas, J.J., Wilson, A.C., Wilhelm, S.: Insight impairment in body image disorders: delusionality and overvalued ideas in anorexia nervosa versus body dysmorphic disorder. Psychiatr. Res. **210**, 1129–1135 (2013)
3. Zhang, A., et al.: Brain connectome modularity in weight-restored anorexia nervosa and body dysmorphic disorder. Psychol. Med. **46**, 2785–2797 (2016)
4. Woodside, B.D., Staab, R.: Management of psychiatric comorbidity in anorexia nervosa and bulimia nervosa. CNS Drugs **20**, 655–663 (2006)

5. Grant, J.E., Kim, S.W., Eckert, E.D.: Body dysmorphic disorder in patients with anorexia nervosa: prevalence, clinical features, and delusionality of body image. Int. J. Eat. Disord. **32**, 291–300 (2002)
6. Ruffolo, J.S., Phillips, K.A., Menard, W., Fay, C., Weisberg, R.B.: Comorbidity of body dysmorphic disorder and eating disorders: severity of psychopathology and body image disturbance. Int. J. Eat. Disord. **39**, 11–19 (2006)
7. Kittler, J.E., Menard, W., Phillips, K.A.: Weight concerns in individuals with body dysmorphic disorder. Eat. Behav. **8**, 115–120 (2007)
8. Cororve, M.B., Gleaves, D.H.: Body dysmorphic disorder: a review of conceptualizations, assessment, and treatment strategies. Clin. Psychol. Rev. **21**, 949–970 (2001)
9. Li, W., et al.: Anorexia nervosa and body dysmorphic disorder are associated with abnormalities in processing visual information. Psychol. Med. **45**, 2111–2122 (2015)
10. Li, W., et al.: Aberrant early visual neural activity and brain-behavior relationships in anorexia nervosa and body dysmorphic disorder. Front. Hum. Neurosci. **9**, 301 (2015)
11. Feusner, J.D., et al.: Abnormalities of visual processing and frontostriatal systems in body dysmorphic disorder. Arch. Gen. Psychiatr. **67**, 197–205 (2010)
12. Beucke, J.C., Sepulcre, J., Buhlmann, U., Kathmann, N., Moody, T., Feusner, J.D.: Degree connectivity in body dysmorphic disorder and relationships with obsessive and compulsive symptoms. Eur. Neuropsychopharmacol. **26**, 1657–1666 (2016)
13. Arienzo, D., et al.: Abnormal brain network organization in body dysmorphic disorder. Neuropsychopharmacology **38**, 1130–1139 (2013)
14. Feusner, J.D., et al.: White matter microstructure in body dysmorphic disorder and its clinical correlates. Psychiatr. Res.: Neuroimaging **211**, 132–140 (2013)
15. Brainstorm Consortium: Analysis of shared heritability in common disorders of the brain. Science **360** (2018). https://doi.org/10.1126/science.aap8757
16. American Psychiatric Association: Diagnostic and Statistical Manual of Mental Disorders: DSM-IV-TR. American Psychiatric Publishing Incorporated (2000)
17. Eisen, J.L., et al.: The Brown assessment of beliefs scale: reliability and validity. Am. J. Psychiatr. **155**, 102–108 (1998)
18. Hamilton, M.: Diagnosis and rating of anxiety. Br. J. Psychiatr. **3**, 76–79 (1969)
19. Montgomery, S.A., Asberg, M.: A new depression scale designed to be sensitive to change. Br. J. Psychiatr. **134**, 382–389 (1979)
20. Phillips, K.A., Hollander, E., Rasmussen, S.A., Aronowitz, B.R., DeCaria, C., Goodman, W. K.: A severity rating scale for body dysmorphic disorder: development, reliability, and validity of a modified version of the Yale-Brown Obsessive Compulsive Scale. Psychopharmacol. Bull. **33**, 17–22 (1997)
21. Mazure, C.M., Halmi, K.A., Sunday, S.R., Romano, S.J., Einhorn, A.M.: The Yale-Brown-Cornell eating disorder scale: development, use, reliability and validity. J. Psychiatr. Res. **28**, 425–445 (1994)
22. Moody, T.D., et al.: Functional connectivity for face processing in individuals with body dysmorphic disorder and anorexia nervosa. Psychol. Med. **45**, 3491–3503 (2015)
23. Sporns, O., Chialvo, D., Kaiser, M., Hilgetag, C.: Organization, development and function of complex brain networks. Trends Cogn. Sci. **8**, 418–425 (2004)
24. Rubin, D.B.: Multiple imputation after 18+ years. J. Am. Stat. Assoc. **91**, 473–489 (1996)
25. Rubin, D.B.: Multiple Imputation for Nonresponse in Surveys. Wiley, Hoboken (2004)
26. Insel, T., et al.: Research domain criteria (RDoC): toward a new classification framework for research on mental disorders. Am. J. Psychiatr. **167**, 748–751 (2010)

Author Index

Printed in the United States
By Bookmasters